EMBRYO EXPERIMENTATION

Ethical, legal and social issues

Edited by
- *PETER SINGER* • *HELGA KUHSE*
- *STEPHEN BUCKLE* • *KAREN DAWSON*
- *PASCAL KASIMBA*

Centre for Human Bioethics
Monash University

CAMBRIDGE
UNIVERSITY PRESS

Published by the Press Syndicate of the University of Cambridge
The Pitt Building, Trumpington Street, Cambridge CB2 1RP
40 West 20th Street, New York, NY 10011–4211, USA
10 Stamford Road, Oakleigh, Victoria 3166, Australia

First published 1990
First paperback edition 1993

Printed in Canada

National Library of Australia cataloguing in publication data:

Embryo experimentation.
Includes index.
ISBN 0 521 38359 5.
1. Human embryo – Transplantation – Moral and ethical aspects.
2. Embryology, Human – Research – Moral and ethical aspects.   3. Fertilization
in vitro, Human – Moral and ethical aspects.   I. Singer, Peter, 1946–
174.25

British Library cataloguing in publication data:

Embryo experimentation.
1. Medicine. Research. Use of human embryos. Ethical aspects
I. Singer, Peter 1946–
174′.28
ISBN 0-521-38359-5

Library of Congress cataloguing in publication data:

Embryo experimentation.
1. Fertilization in vitro, Human – Moral and ethical
aspects.   I. Singer, Peter. [DNLM: A. Ethics, Medical.
2. Fertilization in Vitro. WQ 205 E5 33]
RG135.E43   1990   175   89-25220

ISBN 0-521-38359-5 hardback
ISBN 0-521-43588-9 paperback

Contents

Foreword

A few years ago an editorial appeared in the *Journal of Medical Ethics* under the title 'Two concepts of medical ethics'. The writer distinguishes between 'the obligations of a moral nature which govern the practice of medicine' (Gordon Dunstan's definition in the *Dictionary of Medical Ethics*) and 'the *critical study* of moral problems arising in the context of medical practice'. It is a useful working distinction; and we see it at work in this community.

'The obligations of a moral nature which govern the practice of medicine' arise in their acute and ultimate form in the relation between doctor or nurse and patient, at the bedside, in the laboratory. How these obligations are perceived and passed on is largely a matter for the members of the medical professions, however much they may listen to others, be influenced or controlled by them, be those others ethics committees containing members with skills and talents other than those of medical training, or be they legislators who believe themselves to have been charged by the community with the responsibility of controlling medical experimentation or care.

It is with the second sense, 'the *critical study* of moral problems arising in the context of medical practice', that this book, and behind it the Monash University Centre for Human Bioethics, is concerned. Of course they need each other. The Centre for Human Bioethics would have a sterile existence if there were not medical scientists and practitioners in clinics, hospitals and laboratories aware of obligations; and the work of medical scientists and practitioners, as of committees which advise them and of legislators who would control them, needs the analytical, critical scrutiny of those with minds trained for that purpose. Not surprisingly, and entirely appropriately, that critical study in our community is based in a university. This does not mean that what is said in a volume like this at every point corrects perceptions of obligations obtained elsewhere. After all, even members of the staff of a university only spend a small fraction of their time correcting examination scripts. Their main task, and the task undertaken by the contributors to this symposium, is to ask questions, to ask them critically, that is to say analytically. We do not go to a volume like this to have prejudices, those of ourselves or other people, reaffirmed; but to gain further insight, to think more clearly. This activity must take place if we are to understand better the strengths and weaknesses of what is elsewhere being said and done, or left unsaid and undone, about experiments on embryos. It is to be hoped that the style of this symposium will take out of that discussion some of the heated emotion in which it is sometimes conducted. We cannot expect that it will remove the statement of fundamental convictions from the debate, nor should we wish it to do so; but if it enables us to see more clearly what we are saying when we invoke phrases about the beginning of life, potentiality and the rest it will have served a useful purpose.

Two concluding comments may perhaps be permitted. First, it would be agreed that a discussion of ethical and legal questions is almost fruitless unless we understand the scientific knowledge and processes involved. Many people seem to be able to come to moral and legislative conclusions on the basis of a sketchy or journalistic knowledge of the facts. That would not be good enough for a book or enquiry of this kind. We must be grateful to Dr Karen Dawson who in the following pages puts forth an understanding of the data in terms which are within the grasp of an intelligent and informed reader.

A second comment: it would be easy to be bemused by the newness of the opportunity which IVF provides for research on embryos. Louise Brown's birth in 1978 is in some circles being given the status of a Copernican revolution! We must be grateful to Professor Singer and Dr Kuhse, and to Professor Hare, for putting current concerns in a wider setting. The human race did not start thinking about bioethical questions just yesterday, and it will not stop tomorrow. Indeed, as Professors Gordon Dunstan and John Mahoney in England have shown, a developmental view of the status of the embryo and the fetus has a long and respectable history in Western thought. Perhaps even a little more history of ethical thought would have improved this volume; but one should not ask for everything.

In his highly provocative book, *The Closing of the American Mind*, Professor Allan Bloom has remarked:

> *We are like ignorant shepherds living on a site where great civilizations once flourished. The shepherds play with the fragments that pop to the surface, having no notion of the beautiful structures of which they were once a part. All that is necessary is a careful excavation to provide them with life-enhancing models. We need history, not to tell us what happened, or to explain the past, but to make the past alive so that it can explain us and make a future possible.*

Dare one say it, the universities still exist to remind us of that civilization. We need history, but we need more than history. We need systematic, reflective and reasoned thought; and we need to undertake it together. In that this volume contributes to that continuing discussion on one important theme, embryo experimentation, it is to be welcomed. May it have many successors.

Davis McCaughey

Acknowledgements

This volume had its genesis in 1985, when the National Health and Medical Research Council awarded a three-year Special Initiative Research Grant to a research team based at the Monash University Centre for Human Bioethics and consisting of Dr Margaret Brumby, Dr Helga Kuhse, Professor John M. Swan and Professor Louis Waller. Also involved in the project from an early stage was the centre's Director, Professor Peter Singer. The grant was to investigate bioethical issues in the use of human fetal tissue, *in vitro* human gametes and embryos.

The aims and methodologies of the research were always interdisciplinary in character, requiring contributions from scientists, lawyers and ethicists working together as a team. Initially Dr Karen Dawson, a geneticist, was employed to carry out research on scientific aspects of the project, Pascal Kasimba to study legal questions, and Lorette Fleming, a philosophy graduate, to conduct research on the ethical issues. Subsequently Ms Fleming left to study at Oxford, and was replaced first by Michaelis Michael, and then after his departure by Dr Stephen Buckle. The team of chief investigators also changed. In 1987 Beth Gaze and Dr John Funder replaced Professor Louis Waller and Dr Margaret Brumby as research supervisors; in 1988, when the original grant expired, a related grant was awarded to Professor Singer, Dr Kuhse and Ms Gaze.

This book is one of the products of these grants. We thank the National Health and Medical Research Council for its support; naturally, the council is not responsible for the views expressed in this volume.

We thank in particular John Swan, Margaret Brumby and Louis Waller for helping to put the original team together and for their important contributions during the first stage of the project; Lorette Fleming and Michaelis Michael for their work; and John Funder and Beth Gaze for their continuing assistance. We are also grateful to a still wider range of people who made contributions through a series of public and private seminars and discussions. Medical research workers active in the field of *in vitro* fertilization and embryo transfer gave generously of their time in informing the research team about the current state of these medical arts. We had numerous valuable discussions with other philosophers, bioethicists, lawyers, health professionals and many others. Members of the Advisory Board of the centre gave us advice on several occasions. There are too many people to thank individually; but without them this book would have been much the poorer. Special thanks are due, however, to Heather Mahamooth, secretary to the Centre for Human Bioethics, for her cheerfully given secretarial assistance.

The following chapters of this book have appeared elsewhere in the legal, philosophical, scientific and medical literatures:

Buckle, S., 'Arguing from potential' first published in *Bioethics* 2 (1988), 227–53.

Buckle, S., 'Biological processes and moral events' first published in *Journal of Medical Ethics* 14.3 (1988), 144–7.

Buckle, S., Dawson, K. and Singer, P., 'The syngamy debate: When precisely does a human life begin?' first published in *Law, Medicine and Health Care* 17 (1989), 174–81.

Dawson, K., 'Fertilization and moral status: A scientific perspective' first published in *Journal* of *Medical Ethics* 13 (1987), 173–7.

Dawson, K., 'Segmentation and moral status *in vivo* and *in vitro*: A scientific perspective' first published in *Bioethics* 2 (1988), 1–14.

Gaze, B. and Dawson, K., 'Who is the subject of IVF research?' first published in *Bioethics* 3 (1989).

Gaze, B. and Kasimba, P., 'Embryo experimentation: The path and problems of legislation in Victoria' utilises material written by P. Kasimba and published as 'Experiments on embryos: Permissions and prohibitions under the *Infertility (Medical Procedures) Act 1984* (Vic.)', *Australian Law Journal* 60 (1986), 657–9.

Kasimba, P., 'Regulating IVF human embryo experimentation: The search for a legal basis' first published in *Australian Law Journal* 62 (1988), 128–38.

Kuhse, H. and Singer, P., 'Individuals, humans and persons' contains material from: H. Kuhse, 'Thinking about destructive embryo experimentation' in K. Dawson and J. Hudson (eds), *Proceedings of the Conference—IVF: The Current Debate* (Centre for Human Bioethics, Monash University, Clayton, Vic., 1987), pp 96–105; and P. Singer and H. Kuhse, 'The ethics of embryo research', *Law, Medicine and Health Care* 14, 3–4 (September 1986), 133–8.

Singer, P. and Dawson, K., 'IVF technology and the argument from potential' first published in *Philosophy and Public Affairs* 17 (1988), 87–104.

We are grateful to the initial publishers or copyright owners who have readily agreed to their reproduction.

P.S., H.K., S.B., K.D., P.K.

Introduction

KAREN DAWSON

In 1978 Louise Brown was born: the first child to have resulted from *in vitro* fertilization (IVF) and embryo transfer (ET). (All technical terms used in this book, whether scientific, legal or philosophical, are defined in the glossary.) Since then IVF clinics have been established in many countries around the world and about 38 000 babies have been born as a result of the new reproductive technology. During the 1980s the range of patients suitable for IVF expanded; instead of being a procedure designed to bypass infertility in women with blocked Fallopian tubes, IVF became a method with broad application in the treatment of both male and female infertility, and a potential research agenda extending into many other fields as well.

These developments in IVF and related research aroused a storm of public controversy. Around the world, governments have established committees of inquiry into the social, legal and ethical impacts of IVF technology. There is very little consensus on what should be permitted and what should be prohibited, or even on the question of whether there should be legislation in this area at all. Very few legislatures have as yet translated into law the recommendations put forward by their various committees, although in several countries there has been legislation enacted dealing with some aspects of IVF. (For a summary, see APPENDIX 1 of this volume.) This book attempts to take the debate about IVF and embryo experimentation beyond the level of polemic. The essays it contains seek to present the scientific information needed for an assessment of the issues, and to scrutinize, more rigorously and more systematically than has been done previously, the scientific, ethical, public policy and legal issues which must be addressed wherever the new reproductive technology exists.

Attempts at *in vitro* fertilization (which literally means 'fertilization in glass') and embryo transfer in animals date back to the late nineteenth century. The procedures were initially developed for the study of maternal effects on the embryo before and after birth, but until the 1950s any claims that IVF had been achieved were met with scepticism. This initial scepticism was justified when it was shown, in 1951, that sperm had to undergo 'capacitation' in order to have the ability to fertilize. The 1960s saw IVF and ET established as techniques used widely in animal breeding. Continued research in reproductive biology since then has led to an enormous body of

Fig. 1 Summary of fertilization

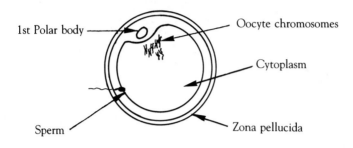

1st Polar body — Oocyte chromosomes

Cytoplasm

Sperm — Zona pellucida

A sperm begins to enter the oocyte

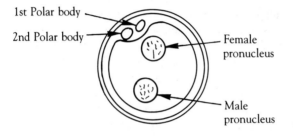

1st Polar body
2nd Polar body — Female pronucleus

Male pronucleus

AFTER 9–12 HOURS

Two pronuclei are clearly visible within the oocyte; one from the sperm and one from the oocyte.

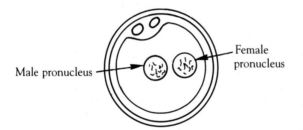

Male pronucleus — Female pronucleus

AFTER 10–22 HOURS

The chromosomes in each pronucleus are drawn together by microtubules in the cytoplasm.

Polar bodies

Zygote chromosomes

22–30 HOURS

The chromosomes of the sperm and oocyte are combined, and syngamy is complete. In 1–3 hours the zygote will undergo the first cleavage division.

information on the *in vitro* requirements for oocyte (egg) maturation, fertiliz-ation and early embryo development for several mammalian species.

This knowledge was first applied to the human in 1965. Robert Edwards published a report on human oocyte maturation *in vitro* and the subsequent progress of this work led to successful *in vitro* fertilization of a human egg. In 1971 Patrick Steptoe, Robert Edwards and Jean Purdy published a de-scription of the first human blastocyst observed after *in vitro* fertilization. This work was the necessary foundation for the first birth from IVF reported by Steptoe and Edwards in 1978.

Knowledge of the process of fertilizing the human oocyte *in vitro* has increased since that initial phase. Fertilization is a complex process lasting for about 24 hours in humans (see Fig. 1), but it is a procedure carried out successfully about 80% of the time in most IVF programmes. Its success continues to be improved by the use of different conditions for fertilization and increasing knowledge of the process of oocyte maturation. The major factors limiting the success of clinical IVF today relate to the optimal con-ditions to ensure continued growth of the embryo after fertilization and the losses which occur after transfer of the embryo into a woman's uterus. PART 1 of this volume contains a discussion of research currently being undertaken to improve IVF and other areas of research which use embryo experimentation.

The scientific research necessary to establish IVF as a clinical process involved the creation of human embryos which were used for research pur-poses. Often they were destroyed in the process of the research. Some proposals for future research also involve the destruction of human embryos. Is such research morally acceptable? If so, under what circumstances? When does the embryo acquire a moral status which is incompatible with its use and destruction in the course of scientific research? When does an individual begin to exist? Is there, at some time during prenatal development, a crucial 'marker event' before which there is no being to whom we have moral obligations, and after which there is? Could this event be fertilization (and is fertilization really an 'event' at all, rather than a process which itself needs to be subdivided)? Could it be a later stage, such as the time at which it ceases to be possible for the pre-embryo to split and become twins? Or later still, when the embryo has become a fetus, and is capable of feeling pain? These questions are to some extent familiar from the debate about abortion; but they need to be reconsidered in the special circumstances created by the existence of an embryo outside a woman's body, and in the light of the new and much more detailed knowledge of embryo development provided by *in vitro* fertilization. So too with the argument that the embryo has a right to protection because, even if it is not a person, it is a potential person: what difference is made by the knowledge we now have about the relatively poor prospects of early embryos ever becoming people?

All these questions are discussed in the essays on ethics in PART 2. The ethical issues are fundamental to the whole public debate, because if we cannot decide whether embryo experimentation is ethically justified (and, if it is, under what conditions) we shall not be in a position to offer a coherent defence of a position either for or against any legislation or regulation which may be proposed.

It has been widely held that resolution of the issues raised by IVF and human embryo research should not be left to the scientists alone; in some instances the scientists themselves have requested government guidelines or legislation to cover some aspects of IVF. Committees set up by governments to consider the new reproductive technology have had to begin by asking how a government should approach such a novel area: should it be by common law, developed by attempting to apply precedents from past cases to novel situations as they come to court? Or by enacting an entirely new piece of legislation? Or through encouraging self-regulation by the scientific and medical professions? The essays in PART 3 of this book deal with the issues that surround the regulation of *in vitro* human embryo experimentation and clinical IVF in a democratic society. Several of them draw on the experience of the Australian state of Victoria, which in 1984 became the first jurisdiction anywhere in the world to adopt comprehensive legislation on IVF. By rushing in where angels feared to tread, the Victorian Government has made mistakes from which others may be able to learn.

The debate about embryo research is far from finished and will continue to occur wherever governments seek to regulate this rapidly developing area of science. The issues discussed in this volume serve to indicate the problems which face any government setting out along this road.

PART 1

THE SCIENTIFIC ISSUES

1 | Introduction: An outline of scientific aspects of human embryo research

KAREN DAWSON

The *in vitro* fertilization of a human egg made it technically possible for the very early human embryo to be used in research, and initiated controversy about whether such research is morally acceptable. In any debate it is important that those interested be able to argue their case from an informed position. In debate about the ethics of human embryo research, arguing from an informed position requires a background knowledge in science, philosophy and law. Some of the central questions to be considered with regard to scientific aspects of this debate include: Is it necessary to use human embryos in research? Why is this research carried out by the scientists? What is the current state of human embryo research? What potential does embryo research offer the scientist in the investigation of problems not able to be studied previously? What are the present limitations to human embryo research? The chapters in PART 1 of this book are concerned with these questions.

IVF provides a unique opportunity for the study of human reproduction and early development, with far-ranging implications for the treatment of infertility and for other areas of research. There are several major areas of research dependent on embryo experimentation in which our present knowledge of this field is being used and added to. These areas are the development of new approaches to contraception, the increased understanding of the mechanisms and diagnosis of infertility, the improvement of the treatment of infertility by IVF and the prenatal detection of genetic and congenital abnormalities. The progress in and problems arising in each of these areas will be discussed briefly here. Before this, however, it is important that we

be familiar with the stages in human development relevant to this research
and to the surrounding debate. This requires an examination of development
during what is referred to as the 'pre-implantation' embryo stage.

Development of the pre-implantation embryo

The term 'pre-implantation embryo' (or sometimes 'pre-embryo') refers to
the entity which exists during the first 14 days of development after fertilization. At about 1–3 hours after the chromosomes (the genetic material
from the sperm and egg) have come together in a process called syngamy,
the first cleavage division of the pre-embryo occurs (see Fig. 2). This division
results in the formation of two cells, known as blastomeres, from the Greek
blastos, meaning 'germ', and *meros*, 'a part'. The blastomeres form inside the
outer membrane of the egg called the zona pellucida (from the Latin *zona*,
'a girdle', and *pellucidus*, 'transparent'). Successive cleavage divisions follow
at about 18-hourly intervals so that the number of cells making up the pre-
embryo increases from two to four to eight, and so on. Though the number
of cells increases as a result of cleavage, the pre-embryo does not increase
in size. It remains just visible to the naked eye, and the blastomeres produced
at each cycle of division become progressively smaller.

Although the chromosomes of the pre-embryo—23 from the egg and
23 from the sperm—have come together at syngamy, the genes of the pre-
embryo do not begin to function until the 4–6 cell stage.[1] Until this time
pre-embryo behaviour is controlled by substances produced before fertilization that were present in the egg.

Following the completion of the third cycle of division, when the
human pre-embryo consists of eight cells, a process called 'compaction'
occurs: the cells adhere to one another and 1–2 cells are pushed to the inside
of the pre-embryo which then starts to develop as a sphere of cells, still
contained within the zona pellucida. Compaction is an important event;
later in development it is the cells then pushed to the inside which will
contribute to the development of the embryo, while the cells on the outside
will become flattened to form trophoblast cells (*trophe* is Greek for 'nutrition')
which later invade the lining of the uterus during implantation and ultimately
make up the embryo's contribution to the placenta.

It is usually at the 2–8 cell stage that the *in vitro* pre-embryo is transferred to the uterus of a woman for further development. By way of comparison, when conception occurs in the normal way within the human body
(*in vivo*), the embryo during this time exists as a free-floating entity that
passes down one of the Fallopian tubes toward the uterus, entering the uterus
3–4 days after fertilization when it consists of 12–16 cells.[2]

The 12–16 cell stage pre-embryo is known as a morula (from the Latin:
'mulberry') and it maintains this name until about the end of the sixth cycle

Fig. 2 Preimplantation embryo development

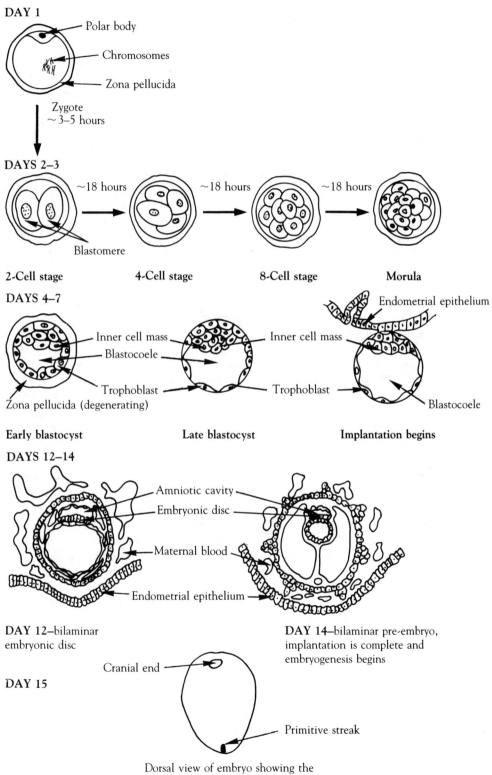

DAY 1

Polar body

Chromosomes

Zona pellucida

Zygote
~3–5 hours

DAYS 2–3

~18 hours   ~18 hours   ~18 hours

Blastomere

2-Cell stage   4-Cell stage   8-Cell stage   Morula

DAYS 4–7

Endometrial epithelium

Inner cell mass   Inner cell mass

Blastocoele

Trophoblast   Trophoblast

Zona pellucida (degenerating)   Blastocoele

Early blastocyst   Late blastocyst   Implantation begins

DAYS 12–14

Amniotic cavity

Embryonic disc

Maternal blood

Endometrial epithelium

DAY 12–bilaminar
embryonic disc

DAY 14–bilaminar pre-embryo,
implantation is complete and
embryogenesis begins

Cranial end

DAY 15

Primitive streak

Dorsal view of embryo showing the
appearance of the primitive streak

of cell division when it consists of 64 cells. Before the 64-cell stage is reached, the zona pellucida surrounding the pre-embryo is degraded by enzymes, a process that is sometimes referred to as 'hatching'. The trophoblast cells then begin to pump fluid into the inside of the sphere of cells, so that the pre-embryo begins to increase in size and become hollow. Reflecting this, the pre-embryo is now known as a blastocyst (from the Greek *blastos*, 'germ', and *kystis* meaning 'bladder') which consists of a fluid-filled cavity called the blastocoele, surrounding the as-yet-undifferentiated inner cell mass which is enclosed within the outer layer of trophoblast cells.

It is now about six days after the completion of fertilization, and the stages described so far (see Fig. 2) are the most relevant to embryo research today. The present technical limitations of *in vitro* cell culture systems lead to difficulties in sustaining *in vitro* pre-embryos in the laboratory for even this length of time. Nevertheless, since the limit of 14 days for *in vitro* embryo research (the completion of the pre-embryo stage of development) has figured in the reports of many committees[3] and is often cited in the debate about embryo research, it is useful to consider development of the pre-embryo up to this time.

About seven days after fertilization the outer cells of the blastocyst begin to invade the lining of the uterus, which is prepared each month to receive a blastocyst and is shed in menstruation if no blastocyst begins to implant. Implantation will not be complete until about 14 days after fertilization when the pre-embryo is totally buried in the lining of the uterus.

The basic structures that will form the embryo (sometimes referred to as the 'embryo proper') begin to become recognizable about 12 days after fertilization. The inner cell mass differentiates into two layers of cells—those which form the embryo itself, and those which form the membrane sacs surrounding the embryo. At this stage the pre-embryo thus consists of two layers of cells; when the pre-embryo stage of development is complete, at about day 15 after fertilization, the embryo will consist of three cell layers. The primitive streak, the first indicator of the embryo's body axis, will then become visible and the development of individual organs from different groups of cells will begin. The pre-embryo is now almost half a millimetre in diameter, about the size of the full stop at the end of this sentence.

To date there has been no documentation of an *in vitro* embryo surviving to this stage. The longest period for which an *in vitro* pre-embryo has been maintained alive is 13 days, but in that instance development did not parallel the normal course of events presented here. The growth of the inner cell mass had ceased and the trophoblast cells had proliferated as out-growths of undifferentiated cells.[4]

Embryo research and contraception

The most widely used methods of contraception are the condom, which provides a physical barrier against the sperm contacting the egg, and the contraceptive pill, which prevents ovulation from occurring in the female. The frequent reports of side-effects associated with prolonged use of the pill prompted further developments in contraceptive research aimed at changing a woman's hormonal balance in more subtle ways—as with the mini-pill[5] —or reducing the fertilizing ability of the male's sperm by the oral administration of substances which impair sperm development or act as spermicides. The success of *in vitro* fertilization with human gametes has, however, made possible other approaches to contraception.

The surface of a sperm is covered with many antigens (substances capable of inducing antibody production); anti-sperm antibodies are found to occur naturally in some cases of female infertility.[6] Some of these sperm antigens are known to be important during fertilization in the progress of sperm in the female tract, its passage through the zona pellucida surrounding the egg and the binding of sperm to the egg membrane. The production of a vaccine of anti-sperm antibodies may provide a new approach to contraception. Before this is possible, however, research is needed to identify and isolate the individual antigens involved.

The use of an antibody vaccine preventing binding of the sperm to the zona pellucida has already been successful in several animal species.[7] These results cannot be directly transposed to the human because of known differences in the interactions between the gametes of different species: the receptors on the zona pellucida involved in sperm penetration and the antigens on the surface of the sperm vary from species to species, although there may be some similarities. (The limitations of using animal models as indicators of the human situation are addressed in more detail in Chapter 2.)

The initial testing of any vaccine developed for human use need not involve embryo research and thus could take place without creating any ethical problems. For example, a vaccine intended to act by inhibiting binding of the sperm to the zona pellucida could be tested using non-viable human eggs, maybe those that have failed to fertilize in IVF or were taken at the time of death from an organ donor. Another possibility for testing is to use salt-stored eggs that have been collected either by laparoscopy or under ultrasound guidance and then treated in a special high salt concentration solution which destroys the egg, but maintains the integrity and functionality of the zona pellucida.[8] Similar testing for a vaccine designed to prevent the sperm binding to the egg membrane could be done using zona-free hamster eggs, a test developed for assessing the fertilizing abilities of human sperm that carries no possibility of an interspecies pre-embryo being formed.[9] (The difficulties of forming interspecies hybrids are considered in Chapter 3 of this book.)

The final test of any vaccine for human use, however, must be carried out on viable human material. In the past this has meant conducting limited clinical trials, with the hope that the contraceptive did not lead to unwanted pregnancies, or abnormalities in any subsequent offspring. Now IVF offers an alternative to the usual clinical trial. The effectiveness of new contraceptives could be assessed by monitoring their ability to block human fertilization *in vitro*. But there are ethical problems with this approach. What if a pre-embryo is inadvertently created during such testing? Is running the risk of this happening justified? The answers to these questions will be important in determining the future course of research into the development of new contraceptives.

Infertility research and embryo experimentation

In addition to being of therapeutic value in infertility treatment, IVF also offers an opportunity for improving the diagnosis of infertility and providing a greater understanding of the mechanisms involved.

The major contribution of IVF in these areas so far has been in the treatment of males diagnosed as clinically infertile with a low sperm count. In some cases, these males are able to father their own children by IVF,[10] indicating that the clinical measure of sperm count is not an absolute indicator of the fertilizing capacity of a male's sperm, and thus of fertility or infertility *per se*. Of the usual clinical tests for assessing the fertility of a male —sperm number, morphology (shape) and motility (movement)—only sperm motility is correlated with fertilizing ability. If a sperm sample contains less than 20% motile sperm, fertilization will not occur.[11]

But if a semen sample contains sperm of adequate numbers, morphology and motility, the male may still be infertile. For fertilization to occur, either *in vitro* or *in vivo*, the sperm must be capable of penetrating the zona pellucida around the egg and of binding to the egg membrane. *In vivo* the sperm must also be capable of passing through the mucus lining the female's cervix. These necessary sperm functions can now be tested in the laboratory.

The ability of sperm to penetrate cervical mucus can be assessed by post-coital examination. (For this reason the test is known as the 'post coital test' or 'PCT'.) The ability of the sperm to pass through cervical mucus correlates highly with successful fertilization *in vitro*,[12] probably because the penetration of the cervical mucus and the zona pellucida both depend on the shearing forces generated by the tail of the sperm as it moves. Because of this, PCT is not only a good indication of sperm function, it also has some predictive value in the outcome of attempts at IVF. Whether or not sperm can bind to the egg membrane can be tested by using the zona-free hamster egg test mentioned above. However, a total picture of sperm function can only be gained if this test is used in conjunction with the PCT. The zona-free hamster egg binding test alone may correlate unreliably with fertilizing ability. Sperm may bind to the egg membrane but be unable to penetrate the zona pellucida.[13] This cause of infertility would be overlooked if only the egg-binding test was used in assessing fertility. The inverse can also occur: sperm may fail to bind to the zona-free hamster egg but successfully fertilize *in vitro* human eggs.[14]

In vitro fertilization allows a more direct and reliable method of testing sperm function. Human eggs donated by women undergoing sterilization operations would provide an accurate assay of sperm function, with the advantage of not requiring the female partner of the male under investigation to undergo an operation for egg collection that may prove to be unnecessary. Ethical problems similar to those raised in relation to contraceptive research also apply here. One difference between the two areas of research, however, is that although any pre-embryo formed as a result of testing in the development of contraceptives is not in itself of major interest when the aim is to prevent fertilization (its presence only indicates failure of the experiment), it is of value in the study of some forms of infertility (for instance, understanding the relationship between sperm abnormalities and early embryo loss may shed light on some types of female infertility). The ethical issues raised by this research concern not only those relevant to the formation of a pre-embryo in the laboratory, but also questions pertaining to how the pre-embryo should be treated and whether it can ethically be used in further study. Is such destructive research to be permitted?

Embryo research in the improvement of IVF

Estimated by the live births occurring, the current success rate for IVF is between 15 and 25% per cycle of egg collection; expressed in a different way, about three out of each hundred eggs collected will result in a live birth.[15] These figures indicate that even though IVF has been successful in leading to the birth of thousands of babies world-wide, there is scope for further improvement. Research in this area has two aims: to improve existing IVF procedures and to introduce new procedures for the treatment of types of infertility that are at present untreatable.

Further improvements in achieving human *in vitro* fertilization can have little influence on the success rate of IVF. The procedure is now fairly routine and is successful about 80% of the time in most laboratories.[16] *In vitro* pre-embryos are usually maintained in the laboratory for about two days following fertilization before transfer to a woman's uterus is considered. During this time, some pre-embryos may develop abnormally, others not develop at all and still others fail to develop to the next stage (a phenomenon known as 'cleavage arrest'). The proportion of *in vitro* pre-embryos leading to a live birth can never be expected to reach 100% of the eggs collected; for even *in vivo* an estimated 60% of fertilized eggs fail to result in a live birth.[17] A success rate more akin to the natural situation would, however, show that embryonic loss from IVF was not an artefact of the procedure.

It is difficult to identify which aspects of the IVF procedure contribute to embryonic loss. The compounds used to induce superovulation for egg collection may be one contributing factor[18] which needs to be further investigated. Such an investigation may not involve embryo experimentation, but could be done by retrospective examination of the treatment used and any effects on the embryo. The superovulants could act either directly on the pre-embryo or indirectly through altering the hormonal environment of the woman. It is partially to avoid any possible complications of the superovulants that cryopreservation of the pre-embryo and the unfertilized egg have been developed. These procedures can be used to allow natural menstrual cycles to recur in the female before embryo transfer is undertaken. Cryopreservation is considered in more detail in Chapter 2.

Information about the nutritional requirements of the human pre-embryo is steadily accumulating, but further controlled studies comparing the effects of different culture media used for maintaining the pre-embryo after fertilization are required to ensure that the medium used provides adequate nutrition and does not affect viability. These experiments could be carried out using spare pre-embryos (not required for transfer to a woman) that have been donated for research, but would perhaps be more efficiently carried out using pre-embryos created solely for research that were derived from donor gametes. The ethical questions raised by this latter approach need to be resolved before such investigations can be contemplated.

The types of infertility treatable by modifications to IVF have continued to increase since its development for the treatment of infertility in females with tubal problems. The most recent example of this is the application of IVF technology for bypassing infertility in the male with oligospermy, where the sperm are weak or few in number and unable to penetrate the zona pellucida of the egg for fertilization. The technique developed involves the injection of a single sperm under the zona pellucida (known as 'sperm microinjection'). The procedure was developed in laboratory tests with human gametes[19] and in animal models. Whether this technique will eventually be transferred to clinical practice depends on the results of present research to assess its safety. Chapter 2, by Alan Trounson, gives a fuller explanation of the procedure.

Embryo research and the detection of genetic abnormalities

It has been estimated that up to 30% of pre-embryos formed by IVF are chromosomally abnormal.[20] This observation does not imply that IVF induces abnormalities (although the possibility of a higher incidence of developmental, as opposed to genetic, abnormalities has been raised[21]). A high frequency of the genetically abnormal *in vitro* pre-embryos would in the *in vivo* situation be lost before pregnancy was detected; the inflated figure for genetic and chromosomal abnormalities in *in vitro* pre-embryos is most likely the outcome of being able to see what would never be detected *in vivo*.

The genetic or chromosomal constitution of any embryo that has led to a pregnancy *in vivo* cannot be diagnosed prenatally before 8 weeks of pregnancy, if chorionic villus sampling (CVS) is used, or 16 weeks if amniocentesis is used. If a defective embryo or fetus is detected, the prospective parents need then to decide whether to terminate or continue the pregnancy. There is at present no way of detecting embryonic abnormalities, especially of a genetic nature, any earlier in pregnancy.

The possibility of detecting abnormalities in the *in vitro* pre-embryo prior to its transfer to the woman is being investigated. A procedure, known as 'embryo biopsy', which involves removing one or two cells from the four- or eight-cell pre-embryo and growing them in culture while the remainder is frozen awaiting the result,[22] is the most likely method for prenatal diagnosis of the pre-embryo. This method, which is explained in more detail in Chapter 2 has proved successful in primates[23] and mouse pre-embryos.[24] For *in vitro* pregnancies, even if termination is an acceptable option to the couple, the decision to terminate is likely to cause even more distress than for a fertile couple. Current research into the detection of genetic and chromosomal disorders may provide an acceptable and less distressing option in the future.

The development of embryo biopsy to the stage that it is used clinically would be of value not only to infertile couples using IVF, where advanced maternal age may increase the risk of a chromosomally abnormal pregnancy (e.g. Down's syndrome), but also to fertile couples in which one or both partners carry a genetic disorder that may be inherited by any offspring. In the long term this technique could decrease the number of therapeutic abortions performed in these two groups. Before embryo biopsy can be used clinically, further research is needed into improving embryo freezing procedures and the development of tests for detecting gene abnormalities. The ethical issues raised by the use of spare pre-embryos and of donor eggs fertilized in the laboratory must also be resolved if any of the above research is to proceed. There are, however, other issues raised by the possible clinical use of embryo biopsy which require consideration by the community. These concern the acceptability of gene therapy. The identification of the genotype of the early embryo opens the future possibility, albeit remote at present, if appropriate techniques can be devised, for any genetic defects found to be corrected before the pre-embryo is transferred to a woman. Although this would enable 'corrected' embryos to be transferred rather than being allowed to succumb in the laboratory, would it then be ethically or legally acceptable to alter the genetic make-up of such pre-embryos? The complexities of this proposal are discussed further in Chapter 3.

Although embryo research has been considered above as four separate areas, the research itself cannot be seen as discrete fields: much useful information in all areas may arise from advances in only one. For instance, improving embryo biopsy may result in an improvement in IVF success rates, etc. In Chapter 2 Alan Trounson, a world leader in IVF research, discusses why scientists involved in IVF see early human embryo research as important. Chapter 3 addresses some of the speculations on future developments which may occur in human reproduction if embryo research continues. Taken together with the background provided here, these essays provide the scientific basis needed for informed consideration of the ethical issues considered in the other chapters of this book.

Notes

1 Braude, P., Bolton, V. and Moore, S., 'Human gene expression first occurs between the four- and eight-cell stage of preimplantation development', *Nature* 332 (1988), 459–61.

2 A comprehensive treatment of embryo development *in vivo* can be found in K. Moore, *The Developing Human*, 3rd ed. (W.B. Saunders, Philadelphia, 1982). For *in vitro* pre-embryo development see S. Fishel, 'Growth of the human conceptus *in vitro*' in S. Fishel and E.M. Edmonds (eds), *In Vitro Fertilization: Past, Present, Future* (IRL Press, Oxford, 1986), pp. 107–26.

3 Reports which have recommended 14 days after fertilization as the limit for embryo research are summarized in L. Walters, 'Ethics and new reproductive technologies: an international review of committee statements', *Hastings Center Report* 17 [3] (1987), Special Supplement 1–9. See also Appendix 1 of this book.

4 Fishel, S.B., Edwards, R.G. and Evans, C.J., 'Human chorionic gonadotropin secreted by preimplantation embryos cultured *in vitro*', *Science* 223 (1980), 816–18.

5 Graham, S. and Fraser, H.M., 'The progesterone-only mini-pill', *Contraception* 26 (1982), 373–88.

6 Jones, H.W. Jr, 'The infertile couple' in Fishel and Symonds, op. cit. (see note 2 above), pp. 17–26.

7 Aitken, R.J. and Lincoln, D.W., 'Human embryo research: the case for contraception', in CIBA Foundation, *Human Embryo Research: Yes or No?* (Tavistock Press, London, 1986), pp. 122–36.

8 Yanagimachi, R., Lopata, A., Odom, C.B., Bronson, R.A., Mahi, L.A. and Nicholson, G.L., 'Retention of biological characteristics of zona pellucidae in highly concentrated salt solution: the use of salt stored eggs for assessing the fertilizing capacity of spermatozoa', *Fertility and Sterility* 31 (1979), 562–74.

9 Yanagimachi, R., Yanagimachi, H. and Rogers, B.J., 'The use of zona free animal ova as a test system for the assessment of the fertilizing capacity of human spermatozoa', *Biology of Reproduction* 15 (1976), 471–6.

10 Mahadevan, M.M., Trounson, A.O., Leeton, J.F. and Wood, C., 'Successful use of *in vitro* fertilization for patients with persisting low-quality semen', *Annals of the New York Academy of Science* 442 (1985), 293–300.

11 Mahadevan, M.M. and Trounson, A.O., 'The influence of seminal characteristics on the success rate of human *in vitro* fertilization', *Fertility and Sterility* 42 (1984), 400–5.

12 Aitken, R.J., 'Andrology and semen preparation for IVF', in Fishel and Symonds, op. cit. (see note 2 above), pp. 89–106.

13 Margalioth, E.J., Navot, D., Lanfer, N., Yosef, S.M., Rabinowitz, R., Yarkoni, S. and Schenker, J.G., 'Zona-free hamster ovum penetration assay as a screening procedure for *in vitro* fertilization', *Fertility and Sterility* 40 (1983), 386–8.

14 Aitken, R.J., 'Andrology and semen preparation for IVF', in Fishel and Symonds, op. cit. (see note 2 above), pp. 89–106.

15 Kola, I., 'Human embryo research: present and future' in K. Dawson and J. Hudson (eds), *Proceedings of the Conference—IVF: The Current Debate* (Centre for Human Bioethics, Monash University, Clayton, Vic., 1987), pp. 12–20.

16 Johnston, I., 'IVF: the Australian experience', submission to the Senate Select Committee on the Human Embryo Experimentation Bill 1985, 26 February 1986, Hansard (AGPS, Canberra, 1986), pp. 560–87.

17 Edmonds, D.K., Lindsay, K.S., Miller, J.F., Williamson, E. and Wood, P.J., 'Early embryonic mortality in women', *Fertility and Sterility* 38 (1982) 447–53.

18 Testarl, J., 'Cleavage stage of human embryos two days after fertilization *in vitro* and their developmental ability after transfer into the uterus', *Human Reproduction* 1 (1986), 29–31.

19 Laws-King, A., Trounson, A., Sathananthan, H. and Kola, I., 'Fertilization of human oocytes by microinjection of a single spermatozoon under the zona pellucida', *Fertility and Sterility* 48 (1987), 637–42.

20 Angell, R.R., Aitken, R.J., Van Look, P.F.A., Lumoden, M.A. and Templeton, A.A., 'Chromosome abnormalities in human embryos after *in vitro* fertilization', *Nature* 303 (1983), 336–8.

21 Lancaster, P.A.L., 'Congenital malformations after *in vitro* fertilization', *Lancet* 2 (1987), 1392–3.

22 Rogers, P.A.W. and Trounson, A., 'IVF: the future' in Fishel and Symonds, op. cit. (see note 2 above), p. 238.

23 McLaren, A., 'Can we diagnose genetic disease in pre-embryos?' *New Scientist* 116 (1987), 42–7.

24 Monk, M., Handyside, A., Hardy, K. and Whittingham, D., 'Preimplantation diagnosis of deficiency of hypoxanthine phosphoribosyl transferase in a mouse model for Lesch-Nyhan syndrome', *Lancet* 2 (1987), 423–5.

2 | Why do research on human pre-embryos?

ALAN TROUNSON

I am frequently asked why we should use human pre-embryos for research.

One of the problems inherent in the debate about embryo experimentation is that of terminology. There is much misunderstanding about what the word 'embryo' actually refers to and what actually constitutes an experiment. A reasonable framework of reference is necessary to address the issues and to come to some meaningful conclusions. The word 'pre-embryo' was introduced by Dr Anne McLaren (a member of the British Government's Warnock Committee and of the United Kingdom Voluntary Licensing Authority) in order to eliminate confusion. It applies to the entity which exists prior to the appearance of the primitive streak at day 14. I shall use this term in the manner recommended by Dr McLaren, in order to distinguish this stage of development from the later embryo.[1]

If you begin with the notion that you shouldn't use pre-embryos unless there is a good argument for doing so, then the reasons will always be very specific. This is essentially how scientists have to approach this type of research because a well-founded case for involving pre-embryos always has to be made to an institutional ethics committee (or even several committees, as in our own situation). In general, scientists have essentially similar views to those of the community in which they are brought up. They consider that the human pre-embryo deserves greater respect than those of other species, in fact they have the same hierarchy of values for pre-embryos of different species as the general community have for fully grown members of different species. Under these circumstances, most research is carried out on embryos of laboratory animals. At a certain point in time, however, no further information can be obtained from laboratory studies. The question then becomes: 'Should this information be used to introduce a new technique, or to alter present techniques being used in clinical medicine?' A case is then developed which depends on the need for the technique, the probability that it will help the medical problem and the risk that another problem will occur. Because of the impossibility of predicting all the differences between species, there is always a risk of unexpected side-effects. The social acceptability of any research project is always a major element in presenting a case for research, and this will be one focus of discussion with ethics committees. The risks have to be evaluated and argued from previous experience, laboratory trials and the likelihood of new problems arising. The potential problems have to be weighed against the potential improvement of the medical problem and on each occasion this will be different.

It is sometimes stated that it is sufficient to cease research at animal studies and not bother with the more difficult ethical problem of studying human tissues or pre-embryos. While cell lines and animal tissues, including gametes and pre-embryos, are useful models to establish new techniques and evaluate the safety of new procedures, it is not always possible to be certain that species-specific effects will not occur. The best known example was the teratogenic effect of thalidomide in humans but not in a wide range of laboratory animals. Sometimes the animal models become inappropriate as well, particularly when the technique has to be modified for human medicine or the animal model doesn't include the specific problem existing in human pathology. Animal and tissue culture models can only be a guide for the likely situation in the human. The use of human pre-embryos requires that some doubt exists whether the animal model can be completely translated to the human situation. The decision will involve an ethical debate about the relative risks and benefits.

An examination of the cases established for two research projects will demonstrate the points mentioned above: the first case concerns the freezing of unfertilized eggs and the second the development of a new fertilization procedure. In the first example, I argued strongly that the risks of serious chromosomal abnormalities arising from the freezing of unfertilized eggs did not warrant the introduction of this technique into clinical IVF. The frozen-thawed human eggs should first be evaluated as pronuclear eggs or pre-embryos before the technique was considered for use clinically. Others did not agree and went ahead and introduced the technique into clinical IVF. The Victorian Government's Standing Review and Advisory Committee on Infertility, to which this case was presented, also disagreed with me and refused my application to study pre-embryos derived from frozen-thawed unfertilized eggs. Many publications have subsequently identified genetic and developmental problems in the freezing of unfertilized eggs.[2] Most of the international IVF community eventually accepted my doubts about the safety of the technique and have ceased its clinical use until further research demonstrates the safety of the procedure. Unless the Victorian Government's Standing Review and Advisory Committee allows proposed research to proceed, this technique will eventually be reintroduced elsewhere in the world, regardless of the risks involved. Our own research group is one of the few with the technical expertise to evaluate adequately the genetic and developmental consequences in animal and human pre-embryos. It would give me no satisfaction if, in retrospect, it was proved that my doubts were justified. Given the difficulty of obtaining approval for testing the safety of the technique in Victoria, our own scientists have decided not to continue research on the cryopreservation of human eggs despite being the international leaders in the field of cell freezing.

The second example is the fertilization of eggs by inserting a sperm under the zona pellucida (shell of the egg). Again I have argued that it is necessary to evaluate the safety of the procedure, using pronuclear human eggs. This was accepted by all our ethics committees and brought about an amendment to the *Infertility (Medical Procedures) Act 1984* (Vic.). (The debate which led to the amendment is discussed in Chapter 19 of this book.) We argued at the time that there was no animal model available to evaluate the safety of the technique, and that sperm from men with poor-quality semen may be more likely to be chromosomally abnormal than sperm from men with better-quality semen. While we were awaiting permission to test the safety of the procedure on pronuclear human eggs, however, we were able to develop an animal model and to demonstrate that the microinjection technique was safe in mice.[3] A majority of the scientists and clinical staff of our own IVF group considered that this new information was a sufficient basis for rearguing the case with the institutional ethics committee. The basis for their argument was that if the technique itself was safe, the patients should decide whether their pronuclear eggs were analysed for chromosomal defects or transferred back to them in order to establish pregnancy. They also felt that there are so many different defects of sperm that no analyses would adequately cover all possibilities of genetic defects arising from such sperm. While I and a minority of other scientists in our group disagreed and wished to analyse eggs before the technique was used clinically, the ethics committee decided to leave the final decision to patients, adding only that the counsellors should advise patients to have some of their pronuclear eggs analysed for chromosomal normality. Since many patients refuse to have any of their pronuclear eggs analysed, we considered there was no further barrier to their treatment.

There is at least an arguable case that the safety of a new technique should under some circumstances involve invasive studies of human pre-embryos and that this may require the fertilization of eggs specifically for this research. Research may also involve pre-embryos donated by patients which are in excess of their needs. The best way to further explore the reasons for pre-embryo research and the possible benefits which may arise from this type of research is to examine some specific examples of recently published reports and a few further possibilities.

Published research involving human pre-embryos

Plachot, M., de Grouchy, J., Junca, A.-M.,
Mandelbaum, J., Salat-Baroux, J. and Cohen, J.,
'Chromosome analysis of human oocytes and embryos: does
delayed fertilization increase chromosome imbalance?' in
Human Reproduction 3 (1988), 125–7.

The aim of this research was to examine chromosomal defects in the human pre-embryo and to determine whether these defects were related to the timing of fertilization. It has been suggested that the ageing egg is more prone to chromosomal defects in natural conception. The researchers analysed the chromosomal normality of 275 pre-embryos at the two- to eight-cell stage. The pre-embryos were donated for research by their patients. Of these, 252 had two pronuclei 17–20 hours after insemination. These were classified as 'timely fertilized'. The other 23 eggs had no pronuclei at 17–20 hours after insemination, but developed pronuclei later and subsequently underwent cleavage. These were classified as 'delayed fertilization'. The chromosome analyses showed that 29.2% of 'timely fertilized' pre-embryos were chromosomally abnormal compared with 87% of 'delayed fertilizaton' pre-embryos.

This study has two important consequences. Firstly, this research shows that about 30% of human pre-embryos have serious chromosomal defects. This is likely to be a major contribution to the failure of pre-embryos to develop to term when replaced in the patient's uterus. The results also confirm earlier studies involving smaller numbers of pre-embryos. Secondly, almost 90% of pre-embryos which do not have pronuclei when expected (17–20 hours after insemination) have serious chromosomal defects. IVF programs normally replace such pre-embryos in the patient's uterus and many IVF clinics reinseminate eggs which do not have pronuclei at 17–20 hours and then replace them if they fertilize after this reinsemination. These 'delayed fertilized' pre-embryos clearly should not be replaced in the patient's uterus if 90% are chromosomally abnormal, nor should they be reinseminated and then replaced *in utero*.

This experiment was important in generating valuable information which may increase the safety of IVF by reducing miscarriage and possible birth defects. In addition it increases our understanding of how birth defects arise from fertilization delay.

Braude, P., Boulton, V. and Moore, S., 'Human gene expression occurs between the four- and eight-cell stage of preimplantation development', Nature 332 (1988), 459–61.

In these studies the authors investigated when the new pre-embryonic genome begins to function. In the first few cleavage divisions the functioning of the cells is controlled by messenger RNA transcribed from the unfertilized egg's DNA (genes). In the mouse, the first *new* protein (not arising from the messenger RNA of the unfertilized egg) is detected at the late two-cell stage. In the pig this occurs at the four-cell stage and in the sheep at the eight-cell stage. The switching on of the genes in development is important for many reasons. For example, it changes the metabolism of the cell and it signals changes in cell development and differentiation. If the new genome does not switch on because of some abnormality, cleavage is likely to cease. Even if only some genes do not switch on, developmental defects are likely, increasing the chance of miscarriage and birth defects.

The authors studied unfertilized eggs, pronuclear eggs and various cleavage stage (spare) pre-embryos up to the blastocyst stage. All of these were incubated briefly (3 hours) in radiolabelled amino acid, and then examined for the proteins which had incorporated radiolabelled amino acids. The proteins were separated on gels in the laboratory and then identified on radiosensitive film. The study showed that pre-embryos up to the four-cell stage produce the same proteins as unfertilized eggs, that is, cell function is controlled by the egg's DNA or messenger RNA. By the eight-cell stage new proteins begin to appear and some of the egg's protein begins to disappear, i.e. the new pre-embryonic genome begins to switch on. The investigators also showed that if the new genome is prevented from switching on by the addition of a chemical which blocks gene transcription, cleavage ceases at the eight-cell stage.

There are a number of implications of this research. Firstly, cleavage and development up to and including the four-cell stage is driven by RNA made by the egg. This means that it will be extraordinarily difficult to identify any factor that will indicate viability of pre-embryos prior to the eight-cell stage; and in fact all attempts to date have been unsuccessful. Cleavage to four-cells is probably unrelated to genetic normality. Secondly, there would be little point in examining pre-embryos prior to the eight-cell stage for any metabolic disorders which have a genetic basis (e.g. Lesch-Nyhan syndrome) by monitoring the products of the gene responsible.[4] In the light of these results, IVF researchers should concentrate on the unfertilized egg or pre-embryo after the eight-cell stage if they are to develop accurate tests of viability using this method. The unfertilized egg, of course, will not have the paternal (sperm) genetic component; hence any test of egg viability will only be 50% accurate. Therefore the preference will be to develop tests of viability at the eight-cell stage or later.

> ### Kola, I., Trounson, A.O., Dawson, G. and Rogers, P., 'Tripronuclear human oocytes: altered cleavage patterns and subsequent karyotypic analysis of embryos', Biology of Reproduction 37 (1987), 395–401.

These studies, carried out at the Monash University Centre for Early Human Development, involved an understanding of the genetic nature of tripronuclear eggs (eggs containing two sperm instead of a single sperm). These eggs cleave and develop to the blastocyst stage *in vitro*, and *in vivo*, to fetal stages and occasionally birth. There has been much discussion about whether or not such eggs are capable of reverting to a normal diploid state (two sets of chromosomes rather than three). We studied eggs with three pronuclei and observed that they usually cleaved directly to three-cells (62% of eggs). When these were analysed we found that all the cells had abnormal numbers of chromosomes which were close to diploidy—that is, they had serious chromosomal abnormalities. Some eggs (14%) did extrude one extra set of chromosomes in a small membrane-bound structure known as a karyoplast and were diploid at the two-cell stage. However, we are uncertain which set of chromosomes was extruded. If a double set of sperm chromosomes was retained, the pre-embryos would be androgenomes which are associated with trophoblastic disease in pregnancy. The remaining eggs (24%) were genuine triploids (three sets of chromosomes) cleaving to two-cells. These may result in miscarriage, trophoblastic disease and very serious birth defects.

As a result of these studies we strongly recommend that tripronucleate eggs be identified at the pronuclear stage and discarded. In the past many IVF clinics did not check eggs for multiple pronuclei and would have replaced many of these in their patients (observed at a frequency of 4–6% of all pronuclear eggs). As a result of this study, it has been widely accepted that this should be done.

> ### Laws-King, A., Trounson, A., Sathananthan, H. and Kola, I., 'Fertilization of human oocytes by microinjection of a single spermatozoon under the zona pellucida', Fertility and Sterility 48 (1987), 395–401.

In this further study at the Centre for Early Human Development, we explored the possibility that a single sperm injected under the zona pellucida would result in the initiation of fertilization and formation of pronuclei. We showed that pronuclei were formed from the injected sperm. This was established by examining the eggs using electronmicroscopy, which found that the remnant of the sperm tail was in the vicinity of one of the pronuclei in each case.

This research raised the possibility of the clinical use of the technique to help infertile couples where the man has severe problems of semen quality. No other treatment is available to such couples except artificial insemination using donor semen.

Other research reported with human pre-embryos has concentrated on the genetic and developmental abnormalities which occur before and at the time of implantation. Similar problems occur in natural conception as demonstrated in data reported from ovum transfer or womb flushing.[5] Whether this research will increase the success rates of IVF or identify some previously unknown barriers to pregnancy cannot be determined at present, but these are the stated aims of the research. At present there are few published studies which are sufficiently well controlled to enable meaningful progress to be made in these areas, because those studies that do exist are usually retrospective analyses of discarded or abnormally developing pre-embryos.

Future prospects of research on human pre-embryos

Up to this point, I have focused mainly on research that has been completed. In the present section I shall discuss some research that is already in progress, as well as other merely possible projects which have been discussed but the conduct of which lies some distance in the future. Some of these projects involve not so much research *on* embryos as the *use of* embryos. At present there are many scientific obstacles which must be overcome before such work can be undertaken. In discussing these possible projects, I am not indicating whether I regard them as ethically acceptable; they will give rise to new ethical questions which must receive careful consideration before any scientific work is done.

The origin and prevention of birth defects

Birth defects arise from genetic and chromosomal anomalies, teratological factors, and errors of developmental embryology. Inheritable genetic disease can be identified in family histories and the likelihood of occurrence for each couple can be predicted. Couples who carry genetic diseases may remain childless, seek to adopt a child, or use donor insemination or egg donation as required. Should they wish to have a child in whom both members have a genetic contribution, however, their options are more limited. Such couples may choose to accept the risk of having a child with the genetic disorder, ignore the possibility, or seek genetic counselling and choose prenatal diagnosis by sampling of chorionic tissue, amniotic cells or fetal tissue or blood. Positive diagnosis of a severe disorder usually results in termination of pregnancy for those who choose prenatal diagnosis. While this is widely accepted in Australia, abortion is an unpleasant choice for any woman to have to make.

Research is underway to develop an alternative which would involve the sampling of the cells from pre-embryos. It has already been shown that the very serious metabolic defect known as Lesch-Nyhan syndrome can be diagnosed in eight-cell mouse pre-embryos and very probably in human pre-embryos.[6] Although metabolic defects of this nature are fairly rare, this success is a beginning for research on this subject. Our own group has concentrated on the possibility of identifying errors of genomic DNA using cell sampling techniques in the mouse. We have explored how the sampling technique affects the development of the pre-embryos and their capacity to be cryopreserved after cell sampling.[7] We have also studied the amplification of DNA of the sample cells, which is necessary for any diagnosis. Using the polymerase chain reaction (PCR) we have shown that it is possible to identify aberrant sequences of DNA in as few as 10 cells from the early cleavage pre-embryo. The possibility exists that cell sampling or biopsy of pre-embryos is an alternative to the prenatal diagnosis of *in utero* embryos or fetuses. The disposal of pre-embryos with serious genetic disorders could be seen as preferable to abortion.

Now that a method for identifying genetic disease in pre-embryos has been established, and if IVF success rates are considered high enough to warrant the use of this technique for prenatal diagnosis, the use of human pre-embryos from patients who have or who carry these disorders should be seriously considered by genetic counsellors and clinicians.

Chromosomal anomalies can already be determined in animal embryos using the cell sampling technique. The incidence of chromosomal anomalies in human pre-embryos is around 30%.[8] These pre-embryos could be identified and discarded if cell sampling techniques were used for all pre-embryos produced in IVF. At the present time it would be difficult to justify the additional costs involved in doing this type of assessment on every pre-embryo produced in IVF. It may, however, be worth considering this technique in association with research aimed to reduce the incidence of chromosomal defects. The relatively high rate of chromosomal anomaly appears to be mostly due to defects in the egg.[9] Hence the initial research strategy should be to reduce the use of eggs with defects. This will involve chromosomal studies on the unfertilized human egg.

Embryo development and the role of proto-oncogenes

There is considerable evidence that proto-oncogenes play a role in cellular proliferation and differentiation. These genes are associated with cancer, transforming cells from a normal cell growth to a malignant state. Some of the proto-oncogenes are expressed during mouse and human embryonic and fetal development, and their activation may be critical in the normal process of embryonic and fetal development.[10] If this is their normal physiological function, their uncontrolled expression at other times may be responsible for malignant cell growth. The search for the origins and factors responsible for cancer has long been a priority in medical research. The control of proto-oncogene expression in embryo development may be a key to these very serious diseases. Research is concentrated at present on animal and cell culture models, and this research may provide some critical information on the role and control of proto-oncogenes in development and where control is lost when cells are transformed at an inappropriate time into malignant growth.

While present research is concentrated on animal embryonic development, a case could be made that human embryos be examined for expression of specific proto-oncogenes in order to gain an understanding of the way in which tumours arise in later development. The examination of pre-embryos donated by patients, embryonic tissue from ectopic pregnancy, or fetal tissue from miscarried or aborted pregnancies could be justified if the potential benefit for the prevention and/or treatment of cancer were substantial.

Grafting of embryonic tissues

Embryonic stem cells are the cells which give rise to the differentiated or specialized cells required for the development of organs or other tissues. The potential use of these embryonic stem cells for grafting has been discussed in the past.[11] The stem cells could be obtained from the embryo in the early post-implantation stage of development.

One of the major problems in present tissue transplantation which uses tissue obtained from adult donors is graft-versus-host disease which occurs during rejection of donated tissue. Embryonic cells have not gained immune competence and do not initiate graft-versus-host disease. It has recently been shown in research on mice that the intravenous injection of embryonic cells into mice whose haemopoietic system (blood cell production system) had been lethally X-irradiated resulted in repopulation of their bone marrow by embryonic haemopoietic stem cells, the production of new red blood cells and lymphocytes and survival of the animals.[12] This research has shown that transplantation of embryonic cells is much more effective than the normal procedure of bone marrow grafting from adult mice, and raises the possibility that the very serious blood diseases such as leukaemia and many anaemias might be treated by intravenous injection of embryonic haemopoietic stem cells without the use of dangerous immune suppression drugs and tissue matching presently necessary. There is still much research required to establish whether embryonic cells can repopulate diseased tissues in otherwise immune-competent animals.

However, the ethics of using embryonic stem cells for treating blood cell diseases in human medicine should be explored. If further research confirms that normal functioning of the haemopoietic system is highly probable, resulting in a complete cure of life-threatening diseases, it is going to be very difficult to ignore the pleas of children and adults for treatment with embryonic stem cells.

One possible source of embryonic stem cells would be ectopic pregnancies. This material is normally discarded. Should this treatment be shown to be effective, however, there would be increasing pressure on scientists to study the early post-implantation embryo in order to provide a readily accessible source of tissue for such treatment.

A *personal perspective*

In his essay, 'Living with Connections', Stephen Jay Gould has written:

> Biologists have rejected as fatally flawed in principle, all attempts by antiabortionists to define an unambiguous 'beginning of life', because we know so well that the sequence from ovulation or spermatogenesis to birth is an unbreakable continuum—and surely no one will define masturbation as murder. Our congressmen may create a legal fiction for statutory effect but they may not seek support from biology.[13]

In fact biological life never begins because sperm and eggs transfer life from one generation to another. The individual's life ceases at death, but the next generation is preserved by the transfer of live cells of the self (gametes) during the reproductive process. There is no sudden cessation of cells of the self and appearance of the new individual. This is a continual process of fusion, cleavage, development, differentiation and growth. These are all essential for the complete evolution of the next individual. Cessation of any component will not enable the process to be completed. My own sensibility is that the individual is complete at birth and rights of protection evolve increasingly with development during pregnancy.

The treatment of infertility, or of debilitating disease or the prevention of such disease, is a value deserving our consideration. The use of preembryonic, embryonic or fetal tissue which would otherwise be disposed of, to treat or to prevent serious disease is certainly worthwhile and consistent with my own values. I am less certain about the deliberate formation of pre-embryos for such research, but I would listen to the arguments and finally decide how my own values can incorporate such a proposal. This is the process of ethics and it is a reasonable way to resolve an issue of conflicting values in the debate about this research in a pluralist society. Creating a rigid legislative framework can only result in mummifying scientific research and associated technology so that progress to improve, simplify and ensure the safety of techniques for overcoming infertility or for treating disease is severely restricted or even halted.

Notes

1 For further discussion of the use of this term, see Chapter 1.

2 Kola, I., Kirby, C., Shaw, J., Davey, A. and Trounson, A., 'Vitrification of mouse oocytes results in aneuplaid zygotes and malformed fetuses', *Teratology* 38 (1988), 467–74.
Johnson, M.H. and Pickering, S.J., 'The effect of dimethylsulphoxide on the microtubular system of the mouse oocyte', *Development* 100 (1987), 313–24.
Johnson, M.H., Pickering, S.J. and George, M.A., 'The influence of cooling on the properties of the zona pellucida of the mouse oocyte', *Human Reproduction* 3 (1988), 383–7.
Trounson, A. and Kirby, C., 'Problems in the cryopreservation of the unfertilized egg by slow cooling in dimethylsulphoxide', *Fertility and Sterility* (1989) (in press).

3 Mann, J.R., 'Full-term development of mouse eggs fertilized by a spermatozoon microinjected under the zona pellucida', *Biology of Reproduction* 38 (1988), 1077–83.

4 Monk, M., Hardy, K., Handyside, A., and Whittingham, D., 'Preimplantation diagnosis of deficiency of hypoxanthine phosphoribosyl transferase in a mouse model for Lesch-Nyhan syndrome', *Lancet* 2 (1987), 423–5.

5 Sauer, M.V., Bustillo, M., Rodi, I.A., Gorrill, M.J. and Buster, J.E., 'In vivo blastocyst production and ovum yield among fertile women', *Human Reproduction* 2 (1987), 701–3.

6 Monk, M. et al., op. cit.

7 Wilton, L.J. and Trounson, A.O., 'Biopsy of preimplantation mouse embryos: cryopreservation and development of micromanipulated embryos and proliferation of single blastomeres *in vitro*', *Biology of Reproduction* 40 (1987), 145–52.

8 See, for example, Angell, R.R., Aitken, R.J., Van Look, P.F.A., Lumsden, M.A. and Templeton, A.A., 'Chromosome abnormalities in human embryos *in vitro* fertilization', *Nature* 303 (1983), 336; Angell, R.R., Templeton, A.A. and Aitken, R.J., 'Chromosome studies in human *in vitro* fertilization', *Human Genetics* 72 (1986), 333–9; and Plachot, M., de Grouchy, J., Junca, A.-M., Mandelbaum, J., Salat-Baroux, J. and Cohen, J., 'Chromosome analysis of

human oocytes and embryos: does delayed fertilization increase chromosome imbalance?' *Human Reproduction* 3 (1988), 125–7.

9 Martin, R.H., Mahadevan, M.M., Taylor, P.J., Hildebrand, K., Long-Simpson, L., Peterson, D., Yamamoto, J. and Fleetham, J. 'Chromosome analysis of unfertilized human oocytes', *Journal of Reproduction and Fertility* 78 (1986), 663–78.

10 Slamon, D.J. and Cline, M.J., 'Expression of cellular oncogenes during embryonic and fetal development of the mouse', *Proceedings of National Academy Sciences USA* 81 (1984), 7141–5.

11 Edwards, R.G., 'The case for studying human embryos and their constituent tissues *in vitro*', in Edwards, R.G. and Purdy, J.M. (eds), *Human Conception in Vitro* (Academic Press, London, 1982), pp. 371–88.

12 Hollands, P., 'Differentiation and grafting of haemopoietic stem cells from early postimplantation mouse embryos', *Development* 399 (1987), 269–76.

13 Gould, S.J., *The Flamingo's Smile: Reflections in Natural History* (Penguin, London, 1985), pp. 64–77.

3 A scientific examination of some speculations about continuing human pre-embryo research

KAREN DAWSON

How scientifically plausible are fears that in the future people will be bred using genetic engineering or cloning, or that human–non-human hybrids will be created? Speculation that these nightmarish consequences will follow from continuing human pre-embryo experimentation has been fuelled by recent advances in reproductive medicine and genetics.[1] Some information for determining whether these claims represent scientific possibility or fantasy, together with a consideration of whether continuing human pre-embryo experimentation will contribute to their realization, is provided in the following discussion.

Using genetic engineering in human breeding

Genetic engineering involves the transfer of genetic material (DNA) determining a specific trait from one individual to another individual from the same species, or of a different species. Some of the human abilities seen as likely candidates for genetic engineering include athletic ability, strength, beauty and intelligence.[2] What are the prospects of creating a race of specially designed people excelling in one of these characteristics using genetic engineering?

A problem confronting any plan for breeding humans is the time required before any results can be assessed. The time involved will be years, which in itself may act as a deterrent for such schemes. One way to reduce the time required to 'tailor-make' people through genetic engineering would be to use female embryos, derived from adult females and males scoring well for traits other than that to be genetically engineered, as a source of eggs. This is in theory possible because egg cells become identifiable in the female embryo at about six weeks after fertilization.[3] It has been proposed that eggs could be obtained from female embryos and matured in the laboratory before being genetically altered and fertilized *in vitro* by the sperm of chosen males to form the next generation.[4]

The immediate practical problem with this proposal is that it is presently not possible to maintain a normally developing embryo *in vitro* for six weeks. The amount of work needed to overcome the technical difficulties of advancing IVF to this point, and then to allow the eggs obtained from the embryo to be reliably matured in the laboratory, make this proposal fanciful. It will not be considered further here.

Ultimately, the important question is: how feasible is the approach of genetically engineering humans for a specific ability? The genetic manipulation of eggs which would be involved (sometimes referred to as 'germ-line genetic engineering'[5]), should not be confused with present attempts at gene therapy. The aims of the two are quite distinct. Current trials of gene therapy involve the provision of a functional gene to replace a non-functional gene for the treatment of diseases caused by single gene defects (such as some inherited blood diseases[6]). The replacement gene is inserted into the somatic cells (body cells) and it has no effect on the germ cells (sperm or eggs). In contrast to germ-line genetic engineering, this type of gene therapy, known as 'somatic gene therapy', does not alter the inheritance of the descendants of the person treated. Germ-line gene therapy is intended to alter the genetic constitution of specific individuals and their descendants. Germ-line genetic manipulation, however, requires further advances before it can be considered feasible.

Techniques for removing or inactivating precise segments of DNA must be devised to allow germ-line gene therapy to have any chance of success. The common method for inserting genes into animal pre-embryos, even genes from other species to form 'transgenic animals',[7] is to insert additional copies of the gene. With this strategy there is no control over how many copies of a gene are present in a cell, or over the site of incorporation of the transferred gene in the DNA of the recipient cell. The problems of this approach are clearly demonstrated by an experiment in which the gene for rat growth hormone was transferred to mice.[8] Large mice were produced, but there were also unexpected side-effects. The incorporation of the gene in the fertilized egg meant that rat growth hormone was produced by most cells in the body, rather than by only the cells of the pituitary gland at the base of the brain, the usual site of its production. The number of copies of the gene in the cells of these animals varied from none to 35, hence different levels of rat growth hormone were produced by each animal. As the site of insertion of the gene for growth hormone was not controlled, some mice were unable to make certain essential proteins. The site of insertion of the rat growth hormone gene had disrupted other genes. Some of the mice were also infertile, and not all of the offspring of the fertile mice were found to carry the gene for rat growth hormone. Instability of the transferred gene must also be overcome.

The most evident effects of an abnormal number of copies of a gene present in a cell in humans are seen in genetic diseases caused by a change in the number of chromosomes or segments of a chromosome present in each cell. As found for Down's syndrome, which results from the presence of an extra copy of chromosome 21, an increase in the number of copies of genes for specific enzymes leads to an excess of the enzymes being produced, and many adverse effects for the affected individual.[9] As either the inability to control the number of copies of a gene in the cells of the pre-embryo, or the site of its insertion in the DNA, could detrimentally affect the essential maintenance of the cell or the later development of the pre-embryo,[10] some precision in removing specific pieces of DNA is essential before germ-line genetic engineering can be used successfully for human breeding.

The traits seen as likely candidates for genetic engineering—strength, intelligence, athletic ability and beauty—are determined by more than one gene, that is they are 'polygenic'. As yet the exact numbers of genes involved in each trait are unknown. We do have some indication of their probable complexity, however, from past surveys of human genetic disorders which affect intelligence. More than a hundred genes play some role in the determination of intelligence, and similar numbers of genes may also be expected to play a role in the determination of the other traits.

A recent application of molecular genetics to the identification of genes involved in some polygenic traits (traits determined by more than one gene) in tomatoes indicates that in the not-too-distant future it may be possible to localize and isolate specific sequences of DNA and to replace them by site-specific insertion.[11] (It needs to be kept in mind that the tomato has 12 chromosomes, and is genetically much simpler than a human with the usual chromosome number of 46.) The equivalent work in animals is at present concerned mainly with single gene replacement,[12] although the same technique used for locating the genes in the tomato (restriction fragment length polymorphism, or RFLP mapping) has also been used to create a similar map of the entire human chromosome complement.[13] The existence of this map by itself, however, is unlikely to lead to the genetically engineering of humans for specific abilities.

There are no genes for intelligence, athletic ability, or any other ability *per se*. These characteristics of an individual which we can see and measure (in genetic terms, a 'phenotype') are the result of the combined effects of the genetic contribution plus effects of the environment. That a phenotype is determined in this way has important ramifications for any attempt at genetically engineering humans for specific abilities.

One consequence of environment affecting the phenotype is that individuals with the same phenotype may have a different genetic make-up (genotype) and, alternatively, that individuals with the same genotype may show a different phenotype because of the environment. For instance, people with different genotypes may achieve a similar physique through diet and exercise, and genetically identical people, such as identical twins, may have quite different physiques because of differences in their diet and exercise. Prior study in different environments would be required to ascertain which genes are important to the determination of intelligence and to rule out the possibility of transferring those genes which are environmentally sensitive.

The genes which contribute to a specific ability also contribute to many other characteristics, a property referred to as 'pleiotropy'. The primary function of a gene is to specify the blueprint of a protein. The protein produced participates in biochemical pathways, and its efficiency to carry out biochemical reactions contributes to an individual's phenotype. Differences in the structure and make-up of the proteins specified by many genes, in combination with the influence of environment, are responsible for the range of abilities seen among individuals. Taken as a whole, the phenotype of a living individual represents a balance of these factors. If we were to insert several genes thought to be related to high intelligence into a person of average or low intelligence, the pleiotropic nature of genes may make the outcome impossible to predict.

Ideally, before any proposal for human genetic engineering is entertained seriously, the location of all human genes (estimated to be about 100 000[14]) and their functions, as well as some information about the interactions between genes, need to be known. Until this requirement is satisfied, there is no sure way of knowing which segments of DNA to alter to achieve the desired effect. But satisfying these requirements is unlikely to be simple. The possibility of unscrupulous persons deciding to use genetic engineering for human breeding before adequate knowledge becomes available cannot be entirely dismissed, but the chances of them being successful in the near future are remote. Knowing the location of the genes and being able to replace them represents only the tip of the iceberg of problems to be overcome, when compared with the understanding required of the possible effects of genetic engineering on the phenotype.

Given the many limitations to using genetic engineering in human breeding, in the near future these methods will not be useful for the creation of beings who excel at specific abilities. The possibility of continued human pre-embryo experimentation leading to the existence of genetically engineered humans remains remote.

Creating human–non-human hybrids

It has been suggested that, in the future, hybrids derived from the interbreeding of different species (interspecies hybrids) might be used for performing dangerous tasks, or for living in inhospitable environments.[15] The prospect of these hybrids carrying human genes horrifies many people.[16] But before this speculation can be used validly as a reason for stopping human pre-embryo experimentation, we must ask whether human–non-human hybrids can be created and whether it would be possible to establish lines or races of these hybrids.

The most commonly known hybrid animal is the mule, a hybrid between the horse and the donkey. The existence of mules means that fertilization between species is possible. Therefore, the creation of human–non-human hybrids must be considered as a possibility. But can this possibility be exploited to establish lines or races of hybrids?

One of the defining characteristics of a species is that it cannot breed with another species to form viable offspring. 'Viability' means not only the ability to survive, but also the ability to produce progeny. It has been estimated that approximately 1 in 50 000 animals in nature is an interspecies hybrid.[17] In evolutionary terms, hybrid animals represent a loss to the gene pool of a population. Many mechanisms have evolved which prevent hybrids from either developing or successfully breeding.

Interspecies hybrids may fail to develop once the zygote is formed because of immunological incompatibility between the female and the embryo. Similarly, a hybrid embryo transferred to one of the species used to form it may be recognized as foreign by the female's immune system and rejected before pregnancy is established. If the hybrid embryo does develop, however, it will be unable to reproduce. The mule is an example of this latter category. Differences between the 64 chromosomes of the horse and the 62 chromosomes of the donkey lead to problems during gamete formation which render the mule sterile, or at least incapable of producing offspring that will survive to birth.[18]

The creation of permanent lines of human–non-human hybrids by interspecies breeding seems likely to remain impossible. The chance of hybrids being repeatedly produced, as with the breeding of mules, however, cannot be ignored. Whether this approach will be adopted in the future will depend on several factors, including the purpose of creating the hybrids. If, as has been suggested, the hybrids will be used to crew interplanetary spacecraft or to inhabit newly established colonies on other planets,[19] their usefulness would be restricted by their distance from Earth and their lifespan, unless all the equipment necessary for the formation of later generations was to be simultaneously transported. If the hybrids are being bred to carry out dangerous tasks, or to inhabit inhospitable parts of this planet, the breeding system would not be a constraint, but we cannot answer whether it will be exploited. The prevailing ethical and social attitudes of that time would need to be known.

There is another method of creating animals derived from two species which also needs to be considered. Strictly speaking, the animals produced are not hybrids in the true sense of the word. The chromosomes of two species are represented in the animal, but they are not present in the same cell. Animals such as these are known as 'chimeras'. The alarm which surrounded the creation of a goat–sheep chimera, called a 'geep', fuelled ideas that this method might one day be used to tailor-make beings for the same supposed uses as human–non-human hybrids.

The geep was formed by including cells from the blastocysts of a goat and a sheep in a single zona pellucida. The resulting animal showed both goat and sheep characteristics.[20] The gametes produced by the geep were either goat or sheep gametes, according to the type of cells which gave rise to them. A geep cannot reproduce its own kind, and nor could a human–non-human being created in the same way. A further disadvantage of creating chimeras for a specific role is that there is no way to predict which of the cells will contribute to the later development of a particular organ or limb. The appearance and characteristics of a chimera are unknown in advance.

Thus, creating a line of human–non-human hybrids or chimeras is probably possible, but not without limitations. Apart from having to re-establish the hybrids or chimeras at each generation (which would take many years if humans were to be used), we cannot predict the physical attributes of the resulting being. As in the case of genetically engineering humans for particular abilities, it is probably not possible to deter unethical people from undertaking the creation of human–non-human hybrids or chimeras for whatever ends, regardless of the risks involved or the small likelihood of success. The techniques necessary for creating human–non-human hybrids and chimeras have been available for years, but have not been used for this purpose. Given this, it is difficult to understand the suggestion that stopping human pre-embryo experimentation would prevent such prospects being realized.

Breeding humans by cloning

An imagined situation for the use of cloning is depicted in the film *The Boys from Brazil*: a dictator wants to produce armies of individuals derived from his cells, who would look like him and think like him. Some people find such a likelihood disturbing, but before these fears are taken seriously as a ground for stopping human pre-embryo experimentation, we need to examine whether these aims can be fulfilled. Can cloning help to accomplish such ambitions?

'Cloning' is a form of asexual reproduction which results in the genetic duplication of an already existing individual. There are two possible methods by which cloning of a human might be achieved: embryo division and nuclear transplantation. Embryo division is used widely in livestock breeding, where the aim is to maximize the number of progeny of a certain genotype or bloodline, and there is no technical reason why the same procedure could not be extended to human use. If human cloning were to be attempted, however, most probably the nucleus from an adult would be used in nuclear transplantation. Briefly, this method of cloning involves removing the nucleus (the part of the cell which contains the chromosomes) from a recently fertilized egg, and replacing it with the nucleus of a cell obtained from the person to be cloned.

As already mentioned, genotype alone does not determine the physical attributes of an individual. The environment begins to influence the phenotype of an individual from the time of fertilization (or maybe even at the time of gamete formation) and continues to do so throughout gestation and after birth. To produce a duplicate of any person would require an exact replication of all aspects of the environments encountered by the individual being cloned at all stages of life. Whether the envisaged armies are intended to be brought into being by transfer of the embryos to surrogate mothers for gestation, or by using the futuristic means of ectogenesis (development outside the uterus) will be irrelevant: the chances of exactly copying the environment experienced by another individual are negligible. Even the possibility of contriving identical environments between the various surrogates used or the different artificial wombs used is remote. Just as identical twins differ physically and in their thoughts and attitudes, so too would members of the same clone.

A further deterrent to the imagined dictator's plan is that the time required for the project is quite formidable. The sort of people envisaged as using cloning are unlikely to be prepared to wait for about 18 years to see their mythical army become a reality. But is the cloning of human beings by nuclear transplantation a technical possibility?

Nuclear transplantation has been used for cloning frogs[21] when the cell cloned is derived from a recently fertilized egg. Intestinal cells taken from tadpoles have also given rise to adult frogs.[22] There are no substantiated claims of success from using an adult donor, although there have been fraudulent claims that mice[23] and an adult human[24] have been cloned by this method. In the scientific literature, there is one unverified report of the initial stages of cloning a human by nuclear transplantation using an undifferentiated nucleus taken from the testes of the donor. The experiment is claimed to have been terminated at the morula stage.[25] Whether it is possible to clone a human who will develop to normal adulthood by nuclear transplantation is unknown, but several pieces of information cast doubt on the possibility.

First, it appears that there are limits to the number of divisions which a mammalian cell can undergo.[26] This accounts for why normal human cells grown in culture eventually senesce, as well as implying that cells taken from an adult would provide unsuitable starting material for cloning. Second, the age of the donor of the nucleus is crucial to the success of cloning. Even cells removed from an eight-cell mouse embryo lead to the successful development of few offspring.[27] As embryo development progresses, cells begin to differentiate, that is to undergo changes which result in later specialization as, for example, a blood cell, a liver cell, etc. Differentiation can involve the loss of genetic material from the cell, as occurs during the development of antibody producing cells, but most often it entails cells undergoing changes which result in a loss of developmental potential, later inherited by their descendants. Differentiated cells usually contain the same genetic material as the zygote from which they arose, but they have lost their ability to become any cell type other than that specified by the changes undergone during differentiation. On this basis, most adult human nuclei are unlikely to be suitable for cloning.

Conclusion

Natural selection, which has acted on hundreds of generations of human ancestors, and which continues to act on us, has resulted in a delicate balance of biological circumstances being required for our survival. It has also culminated in some limits to what actually is biologically possible. There may be some time in the future when these bounds will be overcome. Most likely some of the research necessary will entail pre-embryo experimentation, although it may not necessarily be with human pre-embryos. Whether human breeding using genetic engineering or cloning, or whether the creation of human–non-human hybrids or chimeras will be undertaken then, will depend on the ethical mores at that time, rather than on continuing human pre-embryo experimentation now.

Notes

1 Caton, H., *The Humanist Experiment: Superman from the Test Tube* (Council for a Free Australia, Queensland, 1986).

2 Singer, P. and Wells, D., *The Reproductive Revolution* (Oxford University Press, Oxford, 1984).

3 Moore, K., *The Developing Human*, 3rd ed. (W.B. Saunders, Philadelphia, 1982).

4 Grobstein, C., *Science and the Unborn* (Basic Books, New York, 1988), p. 101.

5 Anderson, W.F., 'Human gene therapy: scientific and ethical considerations', *Journal of Medicine and Philosophy* 10 (1985), 275–91.

6 Williams, D.A. and Orkin, S.H., 'Somatic gene therapy: current status and future prospects', *Journal of Clinical Investigation* 77 (1986), 1053–6.

7 Palmiter, R.D. and Brinster, R.L., 'Transgenic mice', *Cell* 41 (1985), 343–5.

8 Palmiter, R.D. et al., 'Dramatic growth of mice that develop from eggs micro-injected with metallothionein-growth hormone fusion genes', *Nature* 300 (1982), 611–15.

9 Epstein, C.J., 'Aneuploidy in mouse and man' in F. Vogel and K. Sperling (eds), *Human Genetics* (Springer-Verlag, Berlin, 1987), p. 260–68.

10 McNeish, J.D., Scott, W.J. and Potter, S.S., 'Legless: a novel mutation found in PHT1-1 transgenic mice', *Science* 241 (1988), 837–9.

11 Paterson, A.H., Lander, E.S., Hewitt, J.D. et al., 'Resolution of quantitative traits into Mendelian factors by using a complete linkage map of restriction fragment length polymorphisms', *Nature* 335 (1988), 721–6.

12 Doetschman, T., Gregg, R.G., Maeda, N. et al., 'Targetted correction of a mutant HPRT gene in mouse embryonic stem cells', *Nature* 330 (1987), 576–8.

13 Donis-Keller, H., Green, P., Helms, C. et al., 'A genetic linkage map of the human genome', *Cell* 51 (1987), 319–37.

14 McKusick, V.A., 'Human genomics 1986: toward a complete gene map and nucleotide sequence of the human genome' in Vogel and Sperling, op. cit., p. 58.

15 Rorvik, D., *Brave New Baby* (Doubleday, New York, 1971), p. 6.

16 Fletcher, J., *The Ethics of Genetic Control* (Anchor Books, New York, 1974), p. 173.

17 Mayr, E., *Populations, Species and Evolution* (Harvard University Press, Cambridge, Mass., 1970), p. 71.

18 Mettler, L.E. and Gregg, T.G., *Population Genetics and Evolution* (Prentice-Hall, New Jersey, 1969), pp. 193–5.

19 Rorvik, ibid.

20 Fehily, C.B. et al., 'Interspecific chimaerism between sheep and goats', *Nature* 307 (1984), 634–6.

21 Market, C.L., 'Parthenogenesis, homozygosity and cloning in mammals', *Journal of Heredity* 75 (1983), 390–97.

22 Gurdon, J.B. and Uehlinger, V., ' "Fertile" intestine nuclei', *Nature* 210 (1966), 1240–41.

23 Marx, J., 'Three mice cloned in Switzerland', *Science* 211 (1981), 375–6.

24 Rorvik, D., *In His Image: The Cloning of a Man* (Lippincott, Philadelphia, 1978).

25 Shettles, L.B., 'Diploid nuclear replacement in mature human ova with cleavage', *American Journal of Obstetrics and Gynecology* 133 (1979), 222–5.

26 Muggleton-Harris, A.L. and Hayflick, L., 'Cellular aging studied by the reconstruction of replicating cells from nuclei and cytoplasm isolated from normal human diploid cells', *Experimental Cell Research* 103 (1976), 321–30.

27 Tsunoda, Y., Yasui, T., Shioda, Y. et al., 'Full-term development of mouse blastomere nuclei transplanted into enucleated two-cell embryos', *Journal of Experimental Zoology* 242 (1987), 147–51.

PART 2

THE ETHICAL ISSUES

4 | Introduction: The nature of ethical argument

HELGA KUHSE AND PETER SINGER

In the field of embryo experimentation, scientific advance has been followed by ethical controversy from the very beginning. When Robert Edwards and Patrick Steptoe published, in *Nature*, the first account of the fertilization of a human egg outside the body, their experiments were immediately condemned by the Archbishop of Liverpool—and supported by Baroness Summerskill, the social reformer. *The Times* worried that the achievement brought us a step closer to selective breeding; another newspaper raised the spectre of cloning.[1] The ethical debate has continued ever since. The essays which follow take it up once more, and seek to improve upon it.

No one denies that ethical issues are crucial in decision-making about embryo research. Government committees are set up to consider the ethical as well as the social and legal issues raised by the new developments in reproductive technology. The assumption is that only when the ethical issues have been discussed can some form of regulation either be recommended or be seen not to be required. Despite this acceptance of 'ethics', however, an understanding of the nature of ethics, and ethical argument, is often lacking. People still say that ethics is all a matter of subjective opinion. They seem to think that once this has been said, it is obvious that there is rather less scope for serious argument about ethical issues than there is about whether the economy is likely to improve, or even about whether cricket is a better game than baseball. This does not stop them telling us what they take to be right or wrong, but if we do not happen to share the same ethical views, it makes it difficult for the conversation to progress. At the opposite extreme, there are those who maintain that there is no need to discuss ethical questions because the answers are all laid down in a sacred text.

The best way of proving that there is scope for serious argument about ethical issues is to get the arguments going; and that is what the following essays do. Before turning to them, however, it will be helpful to have a brief account of some general points about the nature of ethical argument as understood by many, if not all, contemporary philosophers.

There are some who believe that it is futile to discuss ethics in a pluralistic society, because they think that ethics is always tied to religion, and as long as there is disagreement about religion, ethical discussions will never make progress. Often the basis of this belief lies in the view that the very meaning of 'good' is nothing other than 'what God approves'. Plato refuted a similar claim more than two thousand years ago by arguing that if the gods approve of some actions it must be because those actions are good, in which case it cannot be the gods' approval that makes them good.[2] The alternative view makes divine approval entirely arbitrary: if the gods had happened to approve of torture and disapprove of helping our neighbours, torture would have been good and helping our neighbours bad. If the reply is made that the gods could not approve of torture because they are good and torture is bad, it is apparent that 'good' and 'bad' must have meanings independent of the nature of the gods. This has, indeed, been the dominant view for most of the Judeo-Christian tradition. God has been said to be good, indicating that such ethical judgements are themselves more basic even than religion itself.

If the ethical judgement that something is good or bad is independent of religious belief, can it somehow be based on what it is 'natural' for human beings to do? The problem with this approach is that to describe something as 'natural' is not to state a plain scientific fact but, more often than not, to allow one's values to play a role in judging one kind of behaviour as right and therefore natural, and another kind as wrong and therefore unnatural. The belief that there is one code of behaviour which is uniquely natural to human beings was also put under great strain by the observation of wide variations in behaviour among different cultures on a range of morally sensitive issues, including sexual relations, dietary restrictions, infanticide and euthanasia of the elderly. In any case, even if something were found to be 'natural' in the sense of being present in all human beings, irrespective of cultural influences, it would not necessarily be good. It may be that the use of violence in conflicts for territory, power or status is 'natural' to human beings. That would not mean that we must praise, or even accept, such violence.

Whatever the reason, attempts to ground ethics in human nature, or in what is natural to us, have found little favour in twentieth-century ethics, except for a continuing 'natural law' school among Roman Catholic moral theologians. Recently some sociobiologists, in a burst of enthusiasm about the application of Darwinian evolutionary theory to social behaviour, have talked once more of discovering 'ethical premises' in our biological nature.

So far, however, these ethical premises have turned out to be either vacuous or unconvincing.[3]

But if neither God nor Nature can provide an objective basis for our ethical judgements, are we not, after all, condemned to an ethical relativity based on the standards of our particular society? Or even to purely subjective 'feelings' about what is right and what is wrong? Though there are some philosophers who take this view, the essays which follow show that, at the very least, there are standards of reasoning which prevail in ethics, as in other areas about which we can advance our understanding by discussion and argument.

Perhaps the first elementary point about moral argument is that one has to start with an accurate account of the relevant facts. Just as at law one cannot decide which of several disputing parties is entitled to a share of the lottery prize unless one knows who bought the winning ticket and what arrangements had been made prior to the purchase, so too one cannot discuss the moral significance of different stages of the development of the embryo unless one knows what happened before each stage, and what may happen afterwards. Chapters 5 and 6 by Karen Dawson, on 'Fertilization and moral status' and 'Segmentation and moral status', indicate that some ethical arguments about embryo experimentation should not be accepted because they do not start with an accurate account of the relevant facts. Similarly, Peter Singer and Karen Dawson's 'IVF technology and the argument from potential' (Chapter 8) shows how the situation of an embryo —in particular, whether it is inside a woman's body, or in a glass dish in a laboratory—affects the facts in ways which make a crucial difference to the tenability of traditional arguments about potential.

If getting the facts right is an elementary first point about any argument, so too is the logical requirement of consistency. If I were to argue that all human beings are mortal, and that Socrates was a human being, I would contradict myself if I were to go on to declare that Socrates was immortal. The three assertions I have made would be mutually inconsistent, and hence would fail to affirm anything. We would know that at least one of the claims must be false.

Inconsistency is as fatal a flaw in ethical argument as in any other. As Helga Kuhse and Peter Singer indicate in Chapter 7, 'Individuals, humans and persons', those who oppose abortion make the following pair of assertions:

Every human being has a right to life.
The embryo is a human being.

Once these two claims are accepted, one cannot *consistently* hold:

The embryo has no right to life.

So the argument of opponents of abortion is invincible, *if* the two premises are accepted. Hence the debate often comes down to the acceptability of the premises; that it should do so is silent testimony to the need to avoid inconsistency. Related arguments apply, of course, to embryo experimentation, and are discussed in most of the chapters in PART 2.

Such arguments from consistency often lead, as in the example given, to a discussion of what makes a difference *morally relevant*. How are we to decide? Here a careful attempt to unravel difficult arguments and to clarify obscure or misleading terms can help. Sometimes we are misled or confused by the application of a single term: for instance the term 'human' indubitably applies, in some sense, to the entity which normally arises when a sperm from a human being fertilizes an egg from a human being; but does this mean that such an entity is 'a human being'? If so, what ethical conclusions follow? This point is developed in Chapter 7.

Another tricky notion, often employed in discussions on embryo experimentation, is that of 'potential'. What does it mean to say that an embryo is 'a potential person'? In the abortion debate, the argument from potential is often used as a reason for protecting the fetus from destruction. As already indicated, Chapter 8 questions whether such arguments can be applied—as they often are—to the debate over experimentation on the *in vitro* embryo. In another contribution to this topic (Chapter 9), Stephen Buckle examines ways of 'Arguing from potential' and finds that the argument has often been taken as establishing conclusions which cannot properly be based upon it. He distinguishes some very different notions of 'potential' which need to be sorted out before the argument from potential can be taken further.

The concept of autonomy—the idea of being self-governing—is central to many views of ethics, and hence to many ethical debates. In the case of experimentation on human beings, the freely given consent of the subject is generally regarded an essential safeguard of the subject's autonomy. But in the case of embryo experimentation, as Beth Gaze and Karen Dawson point out, the constant focus on the status of the embryo means that the woman is often overlooked. Hence in Chapter 10 they raise the question: 'Who is the subject of research?' When the position of the woman does come to the fore, as it does in feminist writings, it is often suggested that women patients are under pressure to please those conducting the IVF program, and that this pressure makes it impossible for them to give genuinely autonomous consent to the use of their bodies, and their eggs, for research purposes. In Chapter 11 'Is IVF research a threat to women's autonomy?', Mary Anne Warren addresses this question by seeking to untangle the factual, conceptual and ethical judgements which underlie it.

So far we have seen that in ethics there is scope for careful reasoning in demonstrating inconsistencies in argument, and in learning how to avoid them; and we have also seen that there is a need to examine ethical arguments in some depth and clarify the terms they use. In addition, there is a device often employed by philosophers in ethical arguments which can be helpful in deciding about moral relevance: the imaginary example. A scientist may test a hypothesis by isolating a particular factor in an experiment, and seeing if that factor alone makes a difference. In a related manner, an ethicist may ask us to imagine a case in which every factor is held constant except the one which has been said to make a morally relevant difference. For example, in the real world as far as we know, membership of the human species goes along with a uniquely high capacity for reasoning, for the use of language and for several other characteristics. But is membership of the human species itself important, or is it the other characteristics which matter? We may gain greater insight if we imagine ourselves encountering a friendly, reasoning, communicating Martian and asking what moral significance our attitudes towards such a being would have.

The final, and perhaps most controversial, aspect of ethical argument which we shall mention goes beyond ordinary consistency. It is the requirement that an ethical judgement must be impartial or, in a special sense, universal. This does not mean that only sweeping judgements which one applies in all cases, such as 'never tell a lie', can be ethical. Rather, the essential point is that if I am putting forward a view as an ethical stance, and not merely an expression of my personal interests or feelings, I must be prepared to take a point of view which gives no greater weight to my own interests or attitudes, *simply because they are mine*, than it gives to the interests and attitudes of anyone else affected by the action. This is an ancient idea, to be found in the 'Golden Rule' of the Bible, as well as in the Sanskrit scriptures of the Hindu religion, the teachings of Buddha and the sayings of Confucius. More recently, R.M. Hare has refined the sense of 'universal' required.[4] Ethics demands that we go beyond our individual standpoint, beyond what benefits us and our family or group, and that we adopt a universal point of view. This understanding of ethics is implicit in most, if not all, of the writing on ethics in this volume, and is specifically applied to embryo experimentation in Chapter 7, 'Individuals, humans and persons', as well as in Hare's own essay, 'Public policy in a pluralist society', in PART 3.

The discoveries of Copernicus brought about a revolution in our understanding of our place in the universe. With the cosmos 'all in pieces', as John Donne put it, thinkers like Hobbes and Locke had to grapple anew with the basis of civil society and the nature of political authority. The revolution in reproductive science and technology is similarly compelling us, if in a more limited manner, to work out anew the value of human life at different stages of its existence. The chapters in PART 2 are contributions to that process.

Notes

1 Edwards, R. and Steptoe, P., *A Matter of Life* (Sphere, London, 1981), p. 88.

2 Plato, *Euthyphro*.

3 Examples are: E.O. Wilson, *On Human Nature* (Harvard University Press, Cambridge, Mass., 1978) and R.D. Alexander, *The Biology of Moral Systems* (Aldine de Gruyter, New York, 1987); for discussion, see P. Singer, *The Expanding Circle* (Oxford University Press, Oxford, 1981).

4 See R.M. Hare, *Freedom and Reason* (Oxford University Press, Oxford, 1963) and *Moral Thinking* (Clarendon Press, Oxford, 1981).

5 | Fertilization and moral status: A scientific perspective

KAREN DAWSON

Advances in reproductive technology have made it technically possible for the early human embryo to be an experimental subject. This has enlivened debate concerning the moral status of the prenate,[1] for some consensus on this issue is essential for policy formation aimed at regulating the future of such research.

Within the context of the abortion debate, various landmarks in prenatal development are nominated as the determinant of full moral status. Developmentally the earliest of these is fertilization.[2] Human fertilization is a complex process requiring about 24 hours for completion. Viewed simply, it begins with a spermatozoon, the male gamete, penetrating the ovum or female gamete and culminates in the mingling of the genetic material from each to form a single-celled zygote (see Fig. 1 in the Introduction to this book). This chapter examines some of the arguments for claiming that fertilization is the basis for full moral status in the context of current scientific knowledge. Have these arguments been stated with sufficient precision to cope with the facts as we now understand them? Do they need to be modified and, if so, how might this be done?

Fertilization and moral status: The arguments examined

Arguments in support of fertilization as the time at which full moral status is acquired either rely solely on features of the fertilization process, or on some of its aspects in combination with an emphasis on the potential of the newly formed entity. Those arguments depending on potential are considered in Chapter 8, so the focus here is only on those arguments that rely on features of the fertilization process. For the purposes of discussing the relevant biology, these arguments can be considered as three major types: the genetic argument, the discontinuity/continuity argument and the individuality argument.

The genetic argument

In essence, this approach pin-points fertilization as the time at which moral status is acquired as it is then that entities that 'are genetically human beings'[3] are created. For this argument the crucial event during fertilization is the formation of a human genotype. It is claimed that only at fertilization, and not before, does a new genetic member of the species *Homo sapiens* come about, and at no other point in development is there any 'significant'[4] genetic change. This claim is often coupled with the basic moral principle that it is wrong to destroy innocent human beings which, if taken to include the zygote, leads to the conclusion that it is wrong to destroy early human life from the moment of fertilization.[5]

Biologically, this view raises two major questions— firstly, what constitutes the state of being genetically human, and, secondly, what is meant by a significant genetic change?

Taking the first question: the genome or genetic make-up of an organism may be considered at three levels: the comparatively gross level of the chromosome, the level of the gene itself and the even finer level of molecular structure of the gene. If the condition of being genetically human is considered chromosomally, it is either implicit or explicit[6] in this argument that one prerequisite is the presence of 46 chromosomes (23 of which are contributed by the egg and 23 by the sperm). Presumably these chromosomes are of an accepted karyotypic configuration, for the number of 46 is not unique to humans.[7] A definition of being genetically human based on chromosome number creates problems—for what then is the status of those who fail to fulfil this requirement?

There are numerous human conditions compatible with postnatal life identifiable by the presence of 45, 47 or more chromosomes. The most extreme example results from dispermy: two rather than one sperm enter the egg and participate in fertilization. The result may be the formation of a triploid zygote (a zygote containing 69 or three sets of chromosomes, as opposed to the usual 46 or two sets). Triploidy is estimated to occur in 1–3% of all human fertilizations[8] and *in vitro* as many as 8–10% of fertilizations can be observed to result from the penetration of more than one sperm.[9] The majority of triploids are spontaneously aborted or stillborn but there are some reports of live-born individuals who have lived for up to seven months after birth.[10]

More common conditions showing a variation in chromosome number and accompanied by much longer life-spans include triple-X females (about one in 1500 live female births), Klinefelter's syndrome (about one in 500 live male births) and Down's syndrome (about one in 500 live births) which are usually associated with 47 chromosomes in the karyotype. In contrast to this range of disorders, postnatal existence with only 45 chromosomes is more limited. Turner's syndrome is the only such chromosomal condition in humans, in which one chromosome may be completely absent. The affected females (about one in 4500 live female births) are missing an X-chromosome from the genome.[11]

The incidence of these chromosomal conditions at birth is very much lower than at fertilization, for it is estimated that more than 90% of them are lost through very early pregnancy loss and later spontaneous abortion.[12] The possibility of viable individuals, recognized as human, but with 45, 47 or even more chromosomes, raises a question about the adequacy of using the criterion of chromosome number rigidly, or at least solely, to define a genetic human being. Some further defining characteristics are needed.

Variation at the level of the gene is inherent in the genetic argument and is extremely common in nature. The simplest example of this 'natural variation' is the occurrence of genetic polymorphisms, that is the presence in a population of alternative forms of a gene, known as alleles. This is found for example for the major human blood-typing system, the ABO-blood group, where various combinations of the alleles of this gene occur with varying incidences in different populations.[13]

A similar degree of variation has also become evident at the even finer level of gene structure since molecular genetics provided an extensive array of techniques that permit fragments of the genetic material, DNA, to be isolated and their chemical composition identified. Use of these techniques, especially restriction enzyme mapping,[14] has demonstrated that not all phenotypically equivalent alleles exhibit the same chemical composition in different individuals.

This variation at the genetic level either in the allele present or its chemical composition complicates the definition of an entity as genetically a human being. Where should the boundaries of natural variation be drawn? Can defining a human being solely in genetic terms take account of this range of variation? This type of definition is likely to become increasingly complex as more of the human genome is examined, for at present the genetic content of at most 10% of it is known.[15]

Adoption of the claim of Noonan[16] that '. . . if you are conceived by human parents you are human' may be seen as a possible means of circumventing the difficulties in defining a genetic human being. In the light of recent advances in IVF this approach raises the question: If you are conceived by IVF procedures, from human material, do you still qualify as human? An affirmative answer here surely suggests that the origin of the material rather than any of its characteristics is what is important for specifying a human being. However, even this possible solution may be short-lived, for it relies on species breeding true to their kind. But consider the situation of transgenic animals.

A transgenic animal is one carrying a gene from another species, such as a mouse with a gene from a human source.[17] How many human genes can be introduced into a phenotypic mouse before it is considered as genetically non-mouse or even human? Is it the intra-uterine existence of the entity within a mouse that ensures that such a transgenic animal remains a mouse? Alternatively, would the introduction of a gene from an animal into a human gamete or early embryo, say for the purposes of gene therapy, still ensure that any offspring produced subsequently were human? Even if many animal genes were to be introduced? The advent of transgenic animals blurs the boundaries of the intended meaning of species as 'a discrete breeding unit' and emphasises the need for criteria other than those of the genetic content and its origin, and the site of fertilization being applied in attributing human, and hence moral, status to the prenate. The question 'what is a genetic human being?' does not have a simple answer. From available scientific data it is presently essentially indefinable.

The second part of the genetic argument emphasizes fertilization as the time of acquiring moral status because no further 'significant' genetic changes occur beyond this time. The meaning of 'significant' is not clearly specified in the literature, but genetic changes can and do occur after fertilization—are they significant?

Extreme genetic changes occur during the differentiation of erythrocytes (red blood cells) and cells of the lens of the eye. When finally differentiated these cells contain no nuclear genetic material at all and, conversely, there is a low incidence of cells in the liver that double their genetic content during differentiation.[18] Thus, the postnatal human may be considered as composed of various families of cells some of which have been irreversibly genetically changed during development subsequent to fertilization.

Further genetic changes may also occur throughout the pre- and postnatal phases as a result of randomly occurring heritable changes in the genetic material known as spontaneous mutation. The significance of these random changes depends on their location in the DNA,[19] the stage of development at which they occur[20] and their nature.[21] Mutation is estimated to occur spontaneously at a rate of one per million base pair per round of DNA replication, although this rate differs for specific genes.

It may be argued that the above genetic changes are simply 'variations on a theme' for they are qualitative changes following fertilization, but there are also changes that can occur at this time that alter the genetic make-up of the prenate. These changes result from chromosomal non-disjunction, that is, failure of the chromosomes to separate properly during cell division. The result of non-disjunction during the first division of the zygote is a mosaic organism, that is, an organism composed of two different cell lines, both of which differ genetically from the zygote from which they arose. If non-disjunction occurs during cell division at a later stage of development, it will theoretically result in the formation of an individual with three different cell lines, two of which differ genetically from the zygote. The chromosomal conditions mentioned previously (triple-X, Klinefelter's syndrome, Down's syndrome and Turner's syndrome) may exist either in a mosaic form, resulting from chromosomal non-disjunction following fertilization, or in a pure form from a non-disjunction event during gametogenesis in either parent prior to fertilization. The genetic changes here involve the gain or loss of one whole chromosome after fertilization which may have gross effects on the phenotype, genotype and survival of the developing prenate and should consequently be considered as significant.

In summary, the questions arising from the genetic argument stem mostly from viewing genetic make-up as something static and constant from the moment of fertilization. In reality, differentiation is a dynamic process, and the environment contributes to subtle continuous changes throughout a lifetime. Some acknowledgement of this needs to be made in arguments emphasizing genotype formation as an important aspect of fertilization determining the moral status of the prenate.

The discontinuity–continuity argument

Proponents of this argument[22] view events post-fertilization as comprising a continuum of developmental changes, such that it is impossible to isolate any one stage at which to attribute the attainment of moral status. In contrast to this continuity, fertilization is seen as a radical discontinuity or 'transformation'[23] in development. It is then argued that the union of the two gametes to form the single zygote at fertilization is the only discrete stage at which it can be claimed that a human entity begins to exist.

Emphasis for this interpretation of continuity, known as numerical continuity, is on the change from two gametes to one zygote that is continuous throughout all following development.[24] As previously mentioned in the formation of triploid individuals, more than two entities may sometimes participate in fertilization. A further deviation of this occurs with parthenogenesis—development of the egg without fertilization by a sperm. At present this holds only little relevance to human reproduction, as so far as is known no births have resulted from this process, although the initial stages of parthenogenetic development have been observed rarely *in vitro*.[25] Here, there is no numerical discontinuity unless, as suggested by Quinn,[26] the environmental agent inducing parthenogenetic development is treated as a pre-fertilization entity that is incorporated into the 'zygote' at the onset of development. But this approach implies that environment is irrelevant in the events of normal fertilization, which leads to an arbitrary and unsatisfactory distinction not reflecting the actual course of events.

Dispermy and parthenogenesis represent different deviations in the number of entities participating in the process leading to the initiation of development which may be able to be incorporated into the notion of numerical continuity. But even if these fluctuations are accepted, the notion also specifies that the result of fertilization is the formation of a single entity —the zygote. Is this necessarily the case?

Consider the outcome of dispermy. If dispermy occurs it may predispose to a tumorous condition known as a hydatidiform mole.[27] In such an event embryonic development may not occur. The medical concern with the formation of moles is that they may become malignant and life-threatening for the mother, but here they serve as an example of how fertilization may not necessarily lead to the formation of a zygote.

Now, consider again the outcome of fertilization. The one zygote present in the context of this argument is said to mark the beginning of a human entity that is numerically continuous throughout all subsequent development. Is this the case? What if the zygote should split soon after its formation? Can the notion of numerical continuity cope with this possibility?

Identical twins arise from a single zygote that splits. The mechanism involved is not important here. The point is that with this type of twinning there is only temporary numerical discontinuity, i.e. 1 egg + 1 sperm → 1 zygote → 2 individuals. In trying to make numerical continuity allow for twinning, Quinn[28] concluded that identical twinning, if environmentally determined, was a developmental abnormality. Such an approach is analogous to that used when attempting to reconcile parthenogenesis with numerical continuity, for the nett result is to discount any possible role environment may play in singleton zygote development. At present, the relative contribution of genotype and environment to identical twinning is unclear. Identical twins occur in about one of every 270 pregnancies coming to term,[29] and it has been estimated that a significant proportion of identical twins are lost either through spontaneous abortion[30] or the loss of one fetus which results in singleton development and birth.[31] Further studies of twins during gestation may show identical twins to be more frequent than currently believed. The techniques for such studies to proceed are now becoming available, and if a significant genetic component to identical twin formation could be demonstrated their status, as seen by Quinn, as a 'developmental abnormality' would need reappraisal.

The concept of numerical continuity is too narrow as initially defined to allow for any variations during or subsequent to fertilization. To be ranked as a valid determinant of moral status some refinement of the concept which incorporates current scientific knowledge is needed.

The individuality argument

Proponents of this argument also claim that fundamental moral principles against killing are applicable from fertilization, as this event marks the time when an individual human being begins to exist. The meaning of individual differs among authors. An emphasis on genetics is sometimes integral to the notion—'. . . the unique genetic package of an individual is laid down at fertilization'[32]—while others highlight continuity—'. . . it is the same individual right through from that moment [fertilization] onto the end'.[33]

The problems with these views have largely been discussed. The genotype of an individual later in life is not necessarily that formed at fertilization; many changes can occur subsequently. Similarly, the individual created at fertilization may not remain the same throughout life. The simplest demonstration of this is identical twinning which is possible for about 12 days after fertilization. In this process the original zygote ceases to exist. Conversely, during this time it is also possible for two zygotes derived from the independent fertilization of two eggs to fuse forming a chimera—the one individual resulting from two fertilization events. In neither of these cases is the developing individual the one that formed at fertilization.

Other authors appeal to the potential of the zygote: 'We know that a new human individual organism with the internal potential to develop into an adult . . . comes into existence as a result of the process of fertilization. . .'[34] The intention is not to discuss potential here. However, consideration of this argument does illustrate one feature inherent in many arguments in support of fertilization: a reliance on the viability of the zygote. It is worthwhile noting that for up to 78% of human fertilizations the endpoint is loss[35] rather than progression to the next developmental stage, which can be seen as weakening the applicability of any argument depending on viability.

Summary

Scientific facts alone cannot provide the answer to the debate on the moral status of the prenate. The final outcome is an ethical judgement. However, information about the relevant biological events can create a firmer basis for discussion of such a clearly interdisciplinary issue. Assessment of the arguments claiming that fertilization is the time at which full moral status is acquired is now extremely relevant as policy formation and legislation for the regulation of reproductive technology and pre-embryo research is considered in many countries.

The arguments discussed basically rely on the simple equation that:

$$1 \text{ egg } + 1 \text{ sperm } = 1 \text{ zygote } = 1 \text{ child}$$

This process is seen as marking the beginning of human life, and thus determining that fertilization is the time at which moral status is accorded. The point of whether or not fertilization is the beginning of human life will not be debated here. Leaving this aside, it seems that overall the different arguments for fertilization determining moral status of the prenate oversimplify actual biological events.

Among the problems encountered are an overemphasis on the role of genetics in directing the course of events after fertilization, and a dependence on the fidelity of the new genotype formed at fertilization throughout all subsequent development. Also sometimes inherent in this argument is the assumption that birth will follow from fertilization. Several instances where biology diverges from these claims have been discussed and many more equivalent examples could be cited.

When assessing the claim that fertilization establishes full moral status, several facts should be kept in mind:

- Given suitable environmental conditions, development may sometimes commence without fertilization occurring (parthenogenesis).
- The genotype of any individual may not be that formed at fertilization.
- Development and differentiation after fertilization result in changes to the genetic complement of the prenate.
- Environment is a potent force in the course of development both prenatally and postnatally.
- The formation of a single zygote at fertilization may be the forerunner of the development of multiple identical individuals.
- Successful completion of fertilization in no way assures development through to birth or even the commencement of embryo development.

Until arguments claiming fertilization as the determinant of moral status take into account such facts by being modified to incorporate them, they provide an inadequate basis for policy formation or legislation regulating reproductive technology, as they hold only little relevance to actual biology. Presently, this situation serves to raise the questions of whether the whole issue of moral status needs reappraisal and whether any legislation, either actual or proposed, in this area is premature.

Notes

1 'Prenate', in this chapter, is used as a term for all prenatal stages of development thus reserving the specific terms 'pre-embryo', 'embryo' and 'fetus' for proper usage when applicable.

2 The term 'fertilization' is used throughout in preference to 'conception'. As has been noted elsewhere, 'conception' has become an ambiguous term increasingly associated with implantation, rather than the specific union of the sperm and ovum that is denoted by fertilization. For example, see Federal Republic of Germany, 'Judgement at Karlsruhe' in *Ethical Aspects of Abortion: Some European Views*, IPPF Report, Appendix 2 (1975).

3 Werner, R., 'Abortion: the moral status of the unborn', *Social Theory and Practice* 3 (1974), 201–22.

4 Ibid., 202.

5 R. Werner is the chief proponent of this argument but similar arguments are also provided by, for example, R. Wertheimer, 'Understanding the abortion argument', *Philosophy and Public Affairs* 1 (1971), 67–95; B. Brody, 'On the humanity of the foetus' in T.R. Beauchamp and L. Walters (eds), *Contemporary Issues in Bioethics* (Wadsworth, California, 1978), pp. 229–40; and J. Santamaria, '*In vitro* fertilization and embryo transfer' in M.N. Brumby (ed.), *Proceedings of the Conference—In Vitro Fertilization: Problems and Possibilities* (Centre for Human Bioethics, Monash University, Clayton, Vic., 1982), pp. 48–53.

6 For example, see J.T. Noonan, 'An almost absolute value in history' in J.T. Noonan (ed.), *The Morality of Abortion* (Harvard University Press, Cambridge, Mass., 1970), 1–59.

7 Hsu, T.C. and Benirschke, K., *An Atlas of Mammalian Chromosomes* 10 (Springer-Verlag, New York, 1977).

8 Jacobs, P.A., Angell, R.R., Buchanan, I.M., Hassold, T.J., Matsuyama, A.M. and Mannel, B., 'The origin of human triploids', *Annals of Human Genetics* 42 (1975), 49–57.

9 Plachot, M., Mandelbaum, J., Junca, A., Salat-Baroux, J. and Cohen, J., 'Impairment of human embryo development after abnormal *in vitro* fertilization' in M. Seppälä and R.G. Edwards (eds), *In Vitro Fertilization and Embryo Transfer*, Annals of the New York Academy of Sciences 442 (1985), 336–41.

10 Schinzel, A., *Catalogue of Unbalanced Chromosome Aberrations in Man* (Walter de Gruyter, New York, 1984).

11 Wilson, M.G., 'Cytogenetics update for pediatricians', *Current Problems in Pediatrics* 15 (1985), 1–47.

12 Boué, A., Boué, J. and Gropp, A., 'Cytogenetics of pregnancy wastage', *Advances in Human Genetics* 14 (1985), 1–59.

13 See W.F. Bodmer, and L.L. Cavalli-Sforza, *Genetics, Evolution and Man* (W.H. Freeman, San Francisco, 1976) for a full discussion of racial and population differences.

14 For a general discussion of restriction mapping see R.P. Novick, 'Plasmids', *Scientific American* 243 (1980), 76–83.

15 McKusick, V.A., 'The human gene map', *Clinical Genetics* 27 (1984), 207–39.

16 Jacobs, P.A. et al., op. cit., 51.

17 Palminter, R.D. and Brinster, R.L., 'Transgenic mice', *Cell* 41 (1985), 343–5.

18 Davidson, E.H., *Gene Activity in Early Development* (Academic Press, New York, 1977).

19 Much of the DNA in humans is apparently never used for the production of proteins (the usual gene product) and so spontaneous mutations in this genetic material will have no effect on the phenotype of the organism. Also, the genetic material read for protein production is read as a triplet code and the code is such that any mutation in the third base of a triplet will not always affect the protein

being produced. In contrast, a mutation in the first or second positions of a triplet may have a range of effects on the protein being produced, and may even stop its production.

20 Some genes are only functional at certain stages of development and once that stage is past remain switched off. One instance of this is in the production of human haemoglobin where different clusters of genes are active during the embryo phase, fetal phase and child and adult phases. Mutations during adult life in the embryo phase gene complex would have no effect on the adult, as these genes are then inactive.

21 Mutations are of various types—deletion, insertion, base substitution, inversion and duplication, which may involve only small regions of the genetic material. There are also relatively large-scale structural changes such as transposition, inversion, translocation and duplication which have major effects on the function of the genome.

22 See, for example, G.C. Grisez, *Abortion: The Myths, the Realities and the Arguments* (Corpus Books, New York, 1970); T. Iglesias, 'In vitro fertilization: the major issues', *Journal of Medical Ethics* 10 (1984), 32–7; and Chapter 8 of this book.

23 Quinn, W., 'Abortion: identity and loss', *Philosophy and Public Affairs* 13 (1984), 24–54.

24 See Werner, op. cit., and Quinn, op. cit.

25 Edwards, R. and Trounson, A., 'Discussion on the growth of human embryos *in vitro*' in R.G. Edwards and J.M. Purdy (eds), *Human Conception In Vitro* (Academic Press, London, 1982), 219–33.

26 Quinn, op. cit.

27 Snell, R.S., *Clinical Embryology for Medical Students*, 2nd ed. (Little Brown, London, 1972).

28 Edwards and Trounson, op. cit.

29 Werner, op. cit.

30 Morison, J.E., *Foetal and Neonatal Pathology*, 2nd ed. (Butterworth, London, 1968).

31 Walters, W.A.W. and Renou, P.M., 'Pregnancy care' in C. Wood and A. Trounson (eds), *Clinical In Vitro Fertilization* (Springer-Verlag, New York, 1984), pp. 147–56.

32 Santamaria, J., op. cit. (see note 5 above), p. 49

33 Daly, T., discussion in Brumby, op. cit. (see note 5 above), p. 62.

34 Iglesias, T., op. cit. (see note 22 above), 36.

35 Roberts, C.J. and Lowe, C.R., 'Where have all the conceptions gone?' *Lancet* 1 (1975), 498–9.

6 | Segmentation and moral status: A scientific perspective

KAREN DAWSON

The landmarks in development claimed as determinants of the moral status of the prenate[1] during the abortion debate range from fertilization to birth.[2] In the debate about human embryo research, however, the landmarks in development most highlighted as determinants of moral status are those which occur very early. The present state of IVF technology can only sustain growth of the very early human prenate *in vitro*. As a result, most present embryo research entails the study of fertilization and early cleavage divisions of the zygote.[3]

This emphasis on the very early stages of human development is also evident in reports from several government inquiries established to devise recommendations for the regulation of embryo research. Such inquiries have occurred in several countries,[4] prompted by the concern generated by the use of the human prenate in research. Although no report has recommended that research using the *in vitro*[5] human prenate be permitted beyond 14 days after fertilization, differences among the various inquiries indicate that there is as yet no single landmark in development generally accepted as determining moral status.

Committees which have recommended that legislation be introduced to outlaw all non-therapeutic, or destructive, prenate research[6] subscribe to the view that fertilization determines moral status: from the time of zygote formation at the end of fertilization the prenate acquires a right to life and cannot ethically be destroyed. Other committees, which have recommended that research be permitted for up to 14 days after fertilization,[7] see moral status as being acquired later in development.

The arguments for fertilization as the determinant of moral status have previously been examined and found to have problems in incorporating much present scientific knowledge (see Chapter 5) and only limited applicability to the *in vitro* situation (see Chapter 8). The 14-day limit for prenate research coincides with the end of the initial stage of development now referred to as 'pre-embryo' development.[8] (For further details of pre-embryo development see Fig. 2 in Chapter 1.) During the abortion debate, some writers[9] had also urged that this time was critical for determining moral status because at the completion of this stage there was no longer any possibility of segmentation occurring.

In crude terms, 'segmentation' may be considered as the splitting of the pre-embryo to form multiple identical individuals, or the recombining of twins or triplets to form a single individual (we shall see later the need for this account to be refined).

This chapter is an examination of the arguments in support of the possibility of segmentation as the determinant of moral status, including their relevance to *in vitro* prenates. Can these arguments incorporate the present scientific knowledge of early human development? How applicable is the possibility of segmentation as the determinant of moral status to present *in vitro* research? Consideration of these questions will entail examination of the present applicability of the 14-day limit for *in vitro* embryo research.

Segmentation and moral status: The arguments

Paul Ramsey[10] and Charles Curran[11] claim that the prenate acquires moral status at some time between 6 and 21 days after fertilization, with the specific time nominated differing between authors. The arguments used to substantiate this claim are similar to those used in support of fertilization as the determinant of moral status, in that the concept of individuality and the notion of continuity are common to both arguments. The definition of each concept differs, however, according to the argument being invoked. In arguments for fertilization as the time at which moral status is acquired, the claim is 'that it is the same individual right through from that moment [fertilization] until the end'.[12] Arguments for the importance of segmentation appeal to individual existence not from this moment, but rather from the time beyond which segmentation (be it splitting or recombination) can no longer occur. It is not the individual formed at fertilization that is an entity deserving of moral status: such an individual is present only beyond the time at which twinning or recombination can occur. The concept of individuality is more stringent than in the arguments for fertilization and is applicable only 'when it is irreversibly settled whether there will be one, two or more individuals'.[13]

The notion of continuity is also applied in a more restrictive sense in arguments in support of segmentation. Proponents of this type of argument from fertilization claim that post-fertilization development constitutes a continuum such that no single event after fertilization can be isolated as the time at which moral status is acquired.[14] However, this approach overlooks the possibility that either twinning or recombination may occur in the normal course of events, disrupting the continuum of development. This is overcome in arguments for segmentation by the claim that the continuum of development may properly be thought to 'proceed continuously, without any precise demarcation of the different stages of human life'[15] only after the time at which segmentation is possible has passed.

In biological terms these arguments for segmentation as the determinant of moral status raise two questions for consideration. First, 'segmentation' is poorly defined in most arguments. Different authors have claimed the possibility of segmentation to be the determinant of moral status, but ascribed varying times in development for when segmentation may occur. Is 'segmentation' being used consistently among proponents of this argument? Second, 'irreversible individuality' is central to most of the arguments. The claim is that this state is reached once the possibility of segmentation has passed. How valid is this claim?

'Segmentation': A terminology problem?

'Segmentation' as used medically actually has two meanings: 'division into parts more or less similar' or 'cleavage'.[16] If the reason for nominating segmentation as the determinant of moral status is, as claimed to be, the avoidance of problems caused by post-fertilization events such as twinning and chimera formation (recombination),[17] the arguments in support of segmentation would be expected to depend on the former definition (i.e. division into parts more or less similar). But this seems not always to be so. If segmentation is understood only in this way there is difficulty in reconciling some of the times nominated as important to the determination of moral status with scientific facts.

If 'segmentation' is interpreted under the latter definition of 'cleavage', a different time for the prenate acquiring moral status is obtained. 'Cleavage' is the term used to cover the initial cyclical cell divisions of the pre-embryo, that is, the divisions resulting in the increase of cell number following a 2, 4, 8, 16 pattern. This pattern in division finishes at about six days after fertilization: at this time the zona pellucida of the ovum has disappeared and the blastocyst begins implantation by the trophoblast cells invading the lining of the uterus. An important feature of cleavage is that there is no increase in the size of the pre-embryo throughout subsequent divisions. Without this increase in size implantation cannot commence because the trophoblast cells cannot proliferate sufficiently within the confines of the zona pellucida.

The interpretation of segmentation as meaning cleavage is consistent with the times during development cited by Ramsey, as demonstrated by the following claim:

> *The segmentation of the sphere of developing cells in the case of identical twins (who have the same genotype) . . . occurs by about the time of implantation, that is on the seventh or eighth day after ovulation.*[18]

Similar chronology is also used by Brody in discussion of the different arguments for moral status: 'In a case in which there are identical twins, a primitive streak across the blastocyst signals the separation of the two twins. This occurs about the seventh day after fertilization.'[19] Although specified on different time-scales, the times nominated are equivalent. In the human, fertilization takes about 24 hours and occurs soon after ovulation;[20] therefore Ramsey's claim of 7–8 days after ovulation becomes another way of saying 6–7 days after fertilization.

It is doubtful that the authors intended to use segmentation as meaning 'cleavage', for their arguments would become unable to fulfil their initial intention—to avoid the dilemmas raised by identical twinning and chimera formation. It is true that either of these events may occur during this stage but, as is to be discussed below, the time range in which they are seen to be possibilities is far too narrow.

If segmentation is understood as the alternative definition of 'division into parts more or less similar', the aim of arguments in support of segmentation reverts to the stated intention of avoiding dilemmas posed by twinning and chimera formation. The time at which moral status would be acquired would be when these events ceased to be a biological possibility.

Twinning and higher-order identical multiple births result from the pre-embryo splitting and subsequently developing as more than one prenate. This process is possible at any time up until about 13 or 14 days after fertilization[21] when the primitive streak begins to form in the embryonic disc. This interpretation of segmentation is consistent with the time of 14 days after fertilization as the critical point in development for determining moral status. This time has been nominated by some proponents of the argument for segmentation[22] and is also the point in development cited by others when considering this argument.[23]

The arguments for segmentation are commonly framed in terms of twinning alone rather than both twinning and chimera formation, although several authors do consider both possibilities. The limit assigned for chimera formation seems to be the same as that claimed for twinning.[24]

The formation of a chimera is in many ways the inverse of the twinning process: two or more pre-embryos that have resulted from either a single or independent fertilizations fuse, contributing to the development of a single prenate. Less than twenty cases of human chimeras have been documented.[25]

Whether or not this indicates that chimera formation is actually more un-common than identical twinning, which occurs in about one of each 270 pregnancies,[26] is not really clear. The human chimeras that have been docu-mented have come under medical notice because of gonadal problems, chiefly hermaphroditism, because they are the result of the fusion of a male and female pre-embryo.[27] It could be that chimeras resulting from male–male and female–female fusions occur more frequently but fail to come under medical scrutiny because of normal sexual development and failure to be exposed to medical tests that would indicate their origin. Because of the rarity of recorded cases, the actual upper limit to human chimera formation is unknown. Without this knowledge it does not seem unreasonable to group twinning and chimera formation together with an upper limit of about 14 days.

When the different arguments for segmentation are considered overall, the range of 6–21 days after fertilization is the time at which the segmentation argument is seen to apply to the determination of moral status. This range seems to be unnecessarily extensive. The extension at the lower end of the range (six days after fertilization) may be due to a problem of terminology, segmentation actually being used in two senses. The extension of the rel-evance of the segmentation argument to three weeks after fertilization raises different issues which will be considered below.

Segmentation and irreversible individuality

Although segmentation resulting in the formation of two independent beings may only occur up until the time of primitive streak formation, other events are possible after this time which indicate that the notion of 'irreversible individuality' may need some review if it is to be considered as an important criterion in human life coming 'to be the individual human being it is ever thereafter to be'.[28] There are two conditions which raise questions about the adequacy of this notion: conjoined twins, sometimes known as Siamese twins, and fetus-in-fetu.

Conjoined twins are identical twins that have remained joined rather than forming two separate entities. There are two distinct types of conjoined twins: equal and unequal. Equal conjoined twins consist of two beings at-tached to each other to different extents at variable sites. The attachment may be superficial, involving only skin, or major, with limbs, organs and even brain being in common.[29] Unequal conjoined twins are sometimes known as 'parasitic united twins':[30] one twin is attached to the more perfect twin, most commonly sharing a heart. The rarest and most extreme mani-festation of conjoined twins is known as fetus-in-fetu: the parasitic twin exists within the more perfect twin, usually in the chest or abdominal cavity. This condition is most often recognizable in childhood.[31]

Although conjoined twins and fetus-in-fetu have rarely been documented, the possibility of their occurring raises several points related to the notion of irreversible individuality. Conjoined twins arise from the twinning process occurring after the primitive streak has begun to form, that is, beyond 14 days after fertilization,[32] or, in terms of the argument from segmentation, beyond the time at which irreversible individuality is said to exist. How then are conjoined twins to be classified?

If irreversible individuality is accepted as now understood, conjoined twins must be classified as a single individual (because individuality is assumed to be settled by 14 days after fertilization). But this proposition creates confusion for the sense in which 'individual' is sometimes used in philosophy. Under the previous interpretation the notion of 'irreversible individuality' essentially relies on the usual dictionary definition of individual as a 'discrete entity' or 'something indivisible'. It is doubtful that this is intended. Given that the use of this term is tied up with attributing moral status, and that moral status is conventionally attributed only to human beings, it is more likely and more plausible that 'individual' is being used as a synonym for 'human being' or 'person'.

A person is 'a being that is conscious, in the sense of having the capacity for conscious thought and experiences, but not only that: it must have the capacity for reflective consciousness and self-consciousness . . .'[33] Since pre-embryos do not have these features, 'person' cannot mean the same as 'individual' when applied to the pre-embryo, unless we are prepared to acknowledge that a potential to develop these characteristics is equivalent to their being present already. (There are problems with this definition, including pre-embryos with conditions that result in severe mental retardation, but consideration of these is outside the scope of this chapter.) If the above definition of 'person' is adopted for 'individual', basic to the capabilities required is the possession of a brain, i.e. one individual possesses one brain, two individuals possess two brains, and so on. What does using this definition do to our consideration of conjoined twins? Put simply, it means that conjoined twins are one individual if there is only one brain and two individuals if there are two brains. This situation weakens the possibility of seeing individuality as something irreversibly resolved by about 14 days after fertilization. This in turn raises questions about the adequacy of using the landmark of segmentation in development as the determinant of moral status.

Similar reasoning leads to the same confusion in the case of fetus-in-fetu. This condition results from a combining of identical twins such that one twin is included inside the other at some stage during development so that it shares systems, such as the cardiac system.[34] One case recorded and studied in detail[35] showed that the engulfed twin had developed to the equivalent of four months gestation and consisted of brain, bones, nerve tissue, muscle and some rudimentary organs. Microscopic study showed that engulfment had occurred at about four weeks after fertilization, in terms of the argument for segmentation long after the time when it is claimed that individuality is resolved.

Generally the engulfed twin consists of living tissue and a brain, or the potential for one to develop may or may not be present. If we are to assume that the possession of a brain is integral to the existence of an individual, does this mean that the condition of fetus-in-fetu is considered as affecting two individuals if there are two brains present, but only one if there is only one brain? This variable classification is analogous to that arising when considering conjoined twins. In both instances this confusion is further exacerbated, in terms of the argument for segmentation, by the formation of such an entity after the time at which individuality is seen to be irreversibly resolved.

The possibility of conjoined twins and fetus-in-fetu occurring weakens the applicability of the concept of irreversible individuality, as defined, and similarly the validity of using the proposed argument for segmentation in ascribing moral status from even 14 days after fertilization.

The 14-day limit to in vitro embryo research

Some committee reports about human embryo experimentation[36] have recommended that research using *in vitro* human prenates not be permitted beyond 14 days after fertilization. The recommendation of an upper limit of 14 days for research coincides with segmentation as the determinant of moral status, although not always couched in these terms.

If it is assumed that this upper limit is itself justifiable *in vivo*, to apply this limit justifiably to the *in vitro* situation depends on equivalent rates of development occurring *in vivo* and *in vitro*. There is little knowledge of early human development *in vivo*, but it appears that this may not be so: *in vitro* development may be slower than that *in vivo*.[37] How applicable, then, is the limit of 14 days to *in vitro* embryos? Does it need to be modified?

There are many problems associated with prolonged culture of preembryos, but even if it is successful, the question arises as to whether the collection of cells in culture should actually be considered as a pre-embryo. Further, can the entity in culture for the 14 days properly be said to acquire moral status?

Presently, the creation of *in vitro* human pre-embryos is relatively straightforward, fertilization being successful up to 80% of the time.[38] The comparative simplicity of this procedure is highlighted by the low overall success rate of IVF: only about 10% of successful fertilizations result in live births, although this figure varies between clinics.[39] Once the pre-embryo begins cleavage, the problems of the *in vitro* procedure become evident.

'Cleavage arrest', a state where the pre-embryo remains alive in culture but fails to undergo any further development, may occur at any stage during cleavage.[40] The pre-embryo may then exist in culture for a prolonged period, but its survival is purely temporal and not developmental. (In many respects this can be compared with freezing the pre-embryo.) If a total of 14 days passes with the embryo in cleavage arrest, should it be discarded because it is now deemed under the recommendations to be about to acquire moral status?

The answer to this question depends on whether the 14-day limit is to be interpreted verbatim or whether it is to be construed as the time at which 'irreversible individuality' comes into operation, with formation of the primitive streak being assigned an important moral role. If the former applies the embryo must be discarded, and it becomes obscure as to what the intention of the recommendation really is. If the latter applies, the recommendation can be shown for reasons to be considered below to be irrelevant as a means of regulating present-day embryo research.

Presently, if not transferred to a uterus at the two, four or eight-cell stage, about half of the pre-embryos formed *in vitro* develop successfully to the blastocyst stage. This stage is equivalent to about six days of development *in vivo* when the zona pellucida has degraded and the 'hatched' blastocyst begins to implant in the lining of the uterus.[41] Of the *in vitro* blastocysts surviving to this stage only about 5% go through hatching. For largely unknown reasons this failure to hatch leads to the death of the pre-embryo. One possible cause of death is that the pre-embryo starves from lack of nutrients when left trapped inside the zona pellucida. During this time the zona pellucida undergoes 'tanning', or hardening, by which it becomes chemically altered.[42]

There is difficulty in obtaining growth in culture at or near the time which corresponds with implantation *in vivo*. Very few *in vitro* pre-embryos have been successfully maintained in culture for this long. That two blastocysts have been recorded as surviving in culture beyond this stage—one for 8 days and one for 13 days[43]—may be seen by some to justify the 14-day limit on embryo research at present. This approach, however, oversimplifies things if a parallel with the course of *in vivo* development is still seen as basic to this limit.

The development of the *in vitro* pre-embryo up until about seven days after fertilization is roughly equivalent to that *in vivo*. Beyond this, however, there is no equivalence in development.[44] Development *in vitro* after this time is more equatable with disorganised cancerous growth. Given this, is a limit on the time over which embryo research can be carried out relevant now? No primitive streak will be formed in *in vitro* pre-embryos at this stage; so is this entity really deserving of moral status at present? Can it properly be considered to be a pre-embryo?

Should research manage to rectify these problems, enabling pre-embryos to grow and develop normally in culture for 14 days or beyond, the situation may need to be reconsidered. The relevance of the 14-day limit for embryo research will then depend on the adequacy of arguments and reasoning for holding that this stage marks the time at which moral status is acquired. At present such a limit, which alarms many people as being overgenerous, is in fact misplaced and unnecessary.

Segmentation: The determinant of moral status?

The arguments presented by supporters of segmentation seem to provide an inadequate basis for justifying this stage in development as the determinant of moral status. The apparent use of the two different definitions of segmentation has the effect of artificially inflating the time in development over which these arguments are seen as applicable. To be a strong contender for the determinant of moral status, some clearer definition of segmentation is surely required.

Even if the arguments were to be modified, any case for the time of segmentation to be considered as morally important would still be difficult to defend. This difficulty comes from dependence on the notion of 'irreversible individuality'. The problems are two-fold. Not only can irreversibility be shown not to be applicable in the time period claimed, but also there is a lack of clarity associated with the intended meaning of 'individuality'. It is true that this latter problem is not unique to this argument, but in combination with the former problem it becomes difficult to claim segmentation as morally important from either a scientific or philosophical viewpoint.

Transposing these arguments to the situation of *in vitro* embryo research creates more difficulties. By default more than by substantive reasoning, the argument from segmentation is the basis of legislative recommendations for regulating embryo research. At present the proposed limit of 14 days for *in vitro* embryo research seems unnecessary: technical factors prevent it from becoming an issue. Even if the technical difficulties are resolved, however, it is doubtful whether the entity in culture can properly be considered to attain moral status at this time. The solution to the problems of maintaining growth and of attaining the normal course of development in culture are a long way off. To introduce such control of human embryo research so prematurely seems alarmist and unjustified when based on an inadequate philosophical argument.

Notes

1 'Prenate' is used throughout the chapter as a general term to cover all prenatal stages of development rather than misusing the medically defined terms of

'pre-embryo' (the first two weeks after fertilization), 'embryo' (the start of the third week until the end of the eighth week after fertilization) and 'fetus' (nine weeks after fertilization until birth), which are intended to refer to specific stages of development before birth.

2 Hellegers, A.E., 'Fetal development', *Theological Studies* 31 (1970), 3–9.

3 CIBA Foundation, *Human Embryo Research: Yes or No?* (Tavistock, London, 1986). See also Chapters 1 and 2 of this volume.

4 It is estimated by L. Walters ('Ethics and new reproductive technologies: an international review of committee statements', *Hastings Center Report* (Supplement) 17 (June 1987), pp. 3–9) that at least 25 countries have conducted inquiries into the new reproductive technology, resulting in at least 85 committee statements.

5 Walters, L. op. cit., and Appendix 1 of this volume.

6 For example the Senate Select Committee on the Human Embryo Experimentation Bill 1985, *Human Embryo Experimentation in Australia* (Senator Michael Tate, chairman), (AGPS, Canberra, 1986), p. 51.

7 Committee to Consider the Social, Ethical and Legal Issues Arising from In Vitro Fertilization, *Report on the Disposition of Embryos Produced by In Vitro Fertilization* (Professor Louis Waller, chairman), (Victorian Government Printer, Melbourne, 1984), p. 47, and M. Warnock, *A Question of Life* (Basil Blackwell, Oxford, 1985), p. 66.

8 McLaren, A., 'Where to draw the line?', *Proceedings of the Royal Institute of Great Britain* 56 (1984), 101–20.

9 See, for instance, P. Ramsey, 'Reference points in deciding about abortion' in J.T. Noonan (ed.), *The Morality of Abortion* (Harvard University Press, Cambridge, Mass., 1970), pp. 60–100, and C.E. Curran, 'Abortion: contemporary debate in philosophical and religious ethics' in W.T. Reich (ed.), *Encyclopedia of Bioethics* 1 (The Free Press, London, 1978), pp. 17–26.

10 Ramsey, op. cit. (see note 9 above).

11 Curran, op. cit. (see note 9 above), p. 19.

12 Daly, T., discussion in M.N. Brumby (ed.), *Proceedings of the Conference—In Vitro Fertilization: Problems and Possibilities* (Centre for Human Bioethics, Monash University, Clayton, Vic., 1982), p. 62.

13 Ramsey, op. cit. (see note 9 above), p. 75.

14 Quinn, W., 'Abortion: identity and loss', *Philosophy and Public Affairs* 13 (1970), 24–54.

15 Federal Republic of Germany, 'Judgement at Karlsruhe' in *Ethical Aspects of Abortion: Some European Views*, IPPF Report, Appendix 2, (1975), p. 61.

16 *Dorland's Illustrated Medical Dictionary*, 25th ed. (W.B. Saunders, Philadelphia, 1974).

17 Curran, op. cit. (see note 9 above), p. 19.

18 Ramsey, op. cit. (see note 9 above), pp. 65–6.

19 Brody, B., 'On the humanity of the fetus', in T. Beauchamp and L. Walters (eds), *Contemporary Issues in Bioethics* (Wadsworth, California, 1978), p. 230.

20 Harrison, R.G., *Clinical Embryology* (Academic Press, London, 1978).

21 Morison, J.E., *Foetal and Neonatal Pathology*, 2nd ed. (Butterworth, London, 1968), p. 146.

22 Curran, op. cit. (see note 9 above) and Federal Republic of Germany, op. cit.

23 Hellegers, op. cit. and Federal Republic of Germany, op. cit.

24 Hellegers, op. cit. and Curran, op. cit. (see note 9 above).

25 McLaren, A., *Mammalian Chimeras* (Cambridge University Press, London, 1976) and for a more recent account I.A. Uchida, V.C. Freeman and P. Chen, 'Detection and interpretation of two different cell lines in triploid abortions', *Clinical Genetics* 28 (1985), 489–94.

26 Moore, K., *The Developing Human*, 3rd ed. (W.B. Saunders, Philadelphia, 1982), pp. 129–33.

27 McLaren, op. cit.

28 Ramsey, op. cit. (see note 9 above), p. 66.

29 Morison, op. cit.

30 Willis, R.A., *The Borderland of Embryology and Pathology* (Butterworth, London, 1958), p. 146.

31 Morison, op. cit.

32 Moore, op. cit.

33 Lockwood, M., 'When does a life begin?' in M. Lockwood (ed.), *Moral Dilemmas in Modern Medicine* (Oxford University Press, New York, 1985), p. 10.

34 Sada, I., Shiratori, H. and Nakamura, Y., 'Antenatal diagnosis of fetus in fetu', *Asian-Oceania Journal of Obstetrics and Gynaecology* 12 (1986), 353–6.

35 Yasuda, Y., Mitomori, T., Matsuura, A. and Tanimura, T., 'Fetus-in-fetu: report of a case', *Teratology* 31 (1985), 337–41.

36 Waller, op. cit. and Warnock, op. cit. (see note 7 above).

37 Edwards, R.G., 'Clinical aspects of *in vitro* fertilization' in CIBA Foundation, *Human Embryo Research: Yes or No?* (Tavistock, London, 1986), pp. 39–54.

38 Johnston, I., 'IVF: the Australian experience', submission to the Senate Select Committee on the Human Embryo Experimentation Bill 1985, Hansard, 1986, pp. 560–87.

39 Ibid.

40 Webster, J., 'Embryo replacement' in S. Fishel and E.M. Symonds (eds), *In Vitro Fertilization: Past, Present and Future* (IRL Press, Oxford, 1986), pp. 127–34.

41 Cohen, J., 'Pregnancies, abortion and birth after in vitro fertilization' in Fishel and Symonds, op. cit. (see note 40 above), pp. 135–46.

42 Fishel, S., 'Growth of the human conceptus *in vitro*' in Fishel and Symonds, op. cit. (see note 40 above), pp. 107–26.

43 Fishel, S.B., Edwards, R.G. and Evans, E.J., 'Human chorionic gonadotropin secreted by pre-implantation embryos cultured *in vitro*', *Science* 223 (1980), 816–18.

44 Fishel, op. cit. (see note 42 above).

7 | Individuals, humans and persons: The issue of moral status

HELGA KUHSE AND PETER SINGER

Some well-known arguments against non-therapeutic embryo research rest on the premise that fertilization marks the beginning of 'genetically new human life organized as a distinct entity oriented towards further development',[1] and the additional premise that it is wrong to destroy such life, either because of what it currently is, or because of what it has the potential to become. This type of argument, put forward in the majority report of the Australian Senate Select Committee on the Human Experimentation Bill 1985,[2] can also be found in the Vatican's *Instruction on Respect for Human Life*.[3]

We believe there are good reasons for rejecting this type of argument. In Section I of this chapter we outline the reasons why we should reject the view that a zygote or early human embryo is a distinct human individual;[4] in Section II, we argue against the common view that an early embryo has a right to life because it is an innocent human being; in Section III, we put the argument that experimentation on early human embryos is no harder to justify than various other reproductive choices; and finally, in Section IV, we shall sketch a positive account of how we should think about embryo research.

I

It is often assumed that the answer to the question: 'When does a particular human life begin?' will also provide the answer to the question of how that life ought, morally, to be treated. We shall, however, set the moral question aside for the moment and instead focus on some prior issues that must be faced by anyone who wants to claim that fertilization marks the time when a particular human life or 'I' began to exist.[5]

What this claim amounts to is that the newly fertilized egg, the early embryo and I are, in some sense of the term, the same individual. Now, in one very obvious sense, the zygote that gave rise to me and I, the adult, are not the same individual—the former is a unicellular being totally devoid of consciousness, whereas I am a conscious being consisting of many millions of cells. So the claim that the zygote and I are the same individual must rely on a different sense of 'individual'. And so it does.

It is usually thought that the zygote and I are the same individual in one or both of the following two senses: first, that there is a genetic continuity between the zygote and me (we share the same genetic code); and, second, that there is what, for want of a better term, one might call 'numerical continuity' between us (we are the same single thing). In other words, the zygote does not just have the potential to produce an as-yet-unidentifiable individual, rather the zygote is, from the first moment of its existence, already a particular individual—Tom, or Dick, or Harry. But, as we shall see, this view, which we shall call the 'identity thesis', faces some very serious problems. For, contrary to what is often believed, recent scientific findings do not support the view that fertilization marks the event when a particular, identifiable individual begins to exist.

It is true that the life of the fertilized ovum is a genetically new life in the sense that it is neither genetically nor numerically continuous with the life of the egg or the sperm before fertilization. Before fertilization, there were two genetically distinct entities, the egg and the sperm; now there is only one entity, the fertilized egg or zygote, with a new and unique genetic code. It is also true that the zygote will—other things being equal—develop into an embryo, fetus and baby with the same genetic code.

But, as we shall see, things are not always equal and some serious problems are raised for supporters of the identity thesis.

Here are two scenarios of what might happen during early human development.

In the first scenario, a man and a woman have intercourse, fertilization takes place, and a genetically new zygote, let's call it Tom, is formed. Tom has a specific genetic identity—a genetic blueprint—that will be repeated in every cell once the first cell begins to split, first into two, then into four cells, and so on. On day 8, however, the group of cells which is Tom divides into two separate identical cell groups. These two separate cell groups continue to develop and, some nine months later, identical twins are born. Now, which one, if either of them, is Tom? There are no obvious grounds for thinking of one of the twins as Tom and the other as Not-Tom; the twinning process is quite symmetrical and both twins have the same genetic blueprint as the original Tom. But to suggest that both of them are Tom does, of course, conflict with numerical continuity: there was one zygote and now there are two babies.

People have thought in various ways about this: for example, that when the original cell split, Tom ceased to exist and that two new individuals, Dick and Harry, came into existence. But if that were conceded then it would, of course, no longer be true that the existence of the babies Dick and Harry began at fertilization: their existence did not begin until eight days *after* fertilization. Moreover (and we shall come back to this in a moment) if Tom died on day 8, how is it that he left no earthly remains?

Now consider the second scenario. A man and a woman have intercourse and fertilization takes place. But this time, two eggs are fertilized and two zygotes come into existence—Mary and Jane. The zygotes begin to divide, first into two, then into four cells, and so on. But, then, on day 6, the two embryos combine, forming what is known as a chimera, and continue to develop as a single organism, which will eventually become a baby. Now, who is the baby—Mary or Jane, both Mary and Jane, or somebody else—Nancy?

In one plausible sense of the term, there is genetic continuity between Mary, Jane and the baby. Because the baby is a chimera, she carries the unique genetic code of both Mary and Jane. Some of the millions of cells that make up her body contain the genetic code of Mary, others the genetic code of Jane. So in that sense the baby would seem to be both Mary and Jane. But in terms of numerical continuity, this poses a problem. There is now only one individual where there were formerly two. Does this mean that Mary or Jane, or both of them, have ceased to exist? But to suggest that one of them has ceased to exist poses the problem of explaining why one and not the other should have ceased to exist. Moreover, to say that anyone has ceased to exist will put one in the difficult position (already encountered in the previous example of Tom) of having to explain how it can be that a human individual has ceased to exist when nothing has been lost or has perished—in other words, when there has been a death but there is no corpse.

We could sketch other scenarios to show further complexities, but enough has been said to demonstrate that even before the advent of new reproductive technologies serious problems were raised for the 'identity thesis'. As we shall see, these problems have been compounded by new scientific findings.

It is now believed that early embryonic cells are totipotent; that is, that, contrary to the 'identity thesis', an early human embryo is not one particular individual, but rather has the potential to become one or more different individuals. Up to the 8-cell stage, each single embryonic cell is a distinct entity in the sense that there is no fusion between the individual cells; rather, the embryo is a loose collection of distinct cells, held together by the zona pellucida, the outer membrane of the egg. Animal studies on four-cell embryos indicate that each one of these cells has the potential to produce at least one fetus or baby.

Take a human embryo consisting of four cells. On the assumption that this embryo is a particular human individual, we shall call it Adam. Because each of Adam's four cells is totipotent, any three cells could be removed from the zona pellucida and the remaining cell would still have the potential to develop into a perfect fetus or baby. Now, it might be thought that this baby is Adam, the same baby that would have resulted had all four cells continued to develop jointly. But this poses a problem because we could have left any one of the other three cells in the zona pellucida, each with the potential to develop into a baby. The same baby—Adam? Things are not made any easier by the recognition that the three 'surplus' cells, each placed into an empty zona pellucida, would also have the potential to develop into babies. We now have four distinct human individuals with the potential to develop into four babies. Because it does not make good sense to identify any one particular individual as Adam, let's call them Bill, Charles, David and Eddy.

This example shows that there are not only problems regarding individual identity, but also closely related problems regarding the early embryo's potential to produce one or more human individuals. In the above example, the zygote had the potential to produce either one individual, Adam, or four individuals—Bill, Charles, David and Eddy. But this is not where its potential ends. Had we waited until the embryo had cleaved one more time in its petri dish, there would have been not four, but eight, totipotent cells— that is, eight distinct individual entities oriented towards further development and hence eight potential babies: Fred, Graeme, Harry, Ivan and so on. Moreover, since these individual cells also have the potential to recombine to form, say, just one or two distinct individuals, fertilization cannot be regarded as the beginning of a particular human life.

Those who want to object to embryo experimentation because it destroys a particular or identifiable human life would be on much safer ground were they to argue that a particular human life begins not at fertilization but at around day 14 after fertilization. By that time, totipotency has been lost, and the development of the primitive streak precludes the embryo from becoming two or more different individuals through twinning. Once the primitive streak has formed, it would thus be much easier to argue that it is Adam, Bill or Charles that is developing, or all three of them, but as distinct individuals.

II

Next we want to raise the moral question that we set aside at the beginning of this chapter. Let us assume that we have settled the issue of when a particular individual's life begins—and for the moment it doesn't matter whether this happens at around day 14 or at fertilization. All that we need to assume for our present purposes is that there is such a marker event, that we have identified it and that the entity we are talking about has crossed this particular developmental hurdle.

Now what is it about the new human entity that could raise moral questions about destructive embryo experimentation? Many people believe that it is wrong to use human embryos in research because these embryos are human beings, and all human beings have a right to life. The syllogism goes like this:

Every human being has a right to life.

A human embryo is a human being.

Therefore the human embryo has a right to life.

In case anyone is worrying about issues like capital punishment, or killing in self-defence, we should perhaps add that the term 'innocent' is here and henceforth assumed whenever we are talking of human beings and their rights.

The standard argument has a standard response: to accept the first premise—that all human beings have a right to life—but to deny the second premise, that the human embryo is a human being. This standard response, however, runs into difficulties, because the embryo is clearly a being, of some sort, and it can't possibly be of any other species than *Homo sapiens*. Thus it seems to follow that it must be a human being.

Questioning the first premise

So the standard argument for attributing a right to life to the embryo can withstand the standard response. It is not easy to challenge directly the claim that the embryo is a human being. What the standard argument cannot withstand, however, is a more critical examination of its first premise: that every human being has a right to life. At first glance, this seems the stronger premise. Do we really want to deny that every (innocent) human being has a right to life? Are we about to condone murder? No wonder it is on the second premise that most of the fire has been directed. But the surprising vulnerability of the first premise becomes apparent as soon as we cease to take 'Every human being has a right to life' as an unquestionable moral axiom, and instead inquire into the moral basis for our particular objection to killing human beings.

By 'our particular objection to killing human beings', we mean the objection we have to killing human beings, over and above any objection we have to killing other living beings, such as pigs and cows and dogs and cats, and even trees and lettuces. Why do we think killing human beings is so much more serious than killing these other beings?

The obvious answer is that human beings are different from other animals, and the greater seriousness of killing them is a result of these differences. But which of the many differences between humans and other animals justify such a distinction? Again, the obvious response is that the

morally relevant differences are those based on our superior mental powers—
our self-awareness, our rationality, our moral sense, our autonomy, or some
combination of these. They are the kinds of thing, we are inclined to say,
which make us 'uniquely human'. To be more precise, they are the kinds
of thing which make us persons.

That the particular objection to killing human beings rests on such
qualities is very plausible. To take the most extreme of the differences
between living things, consider a person who is enjoying life, is part of a
network of relationships with other people, is looking forward to what tomorrow may bring, and is freely choosing the course her or his life will take
for the years to come. Now think about a lettuce, which, we can safely
assume, knows and feels nothing at all. One would have to be quite mad,
or morally blind, or warped, not to see that killing the person is far more
serious than killing the lettuce.

We shall postpone asking which mental qualities make it more morally
serious to kill a person than to kill a lettuce. For our immediate purposes,
we will merely note that the plausibility of the assertion that human beings
have a right to life depends on the fact that human beings generally possess
mental qualities that other living beings do not possess. So should we accept
the premise that every human being has a right to life? We may do so, but
only if we bear in mind that by 'human being' here we refer to those beings
who have the mental qualities which generally distinguish members of our
species from members of other species.

If this is the sense in which we can accept the first premise, however,
what of the second premise? It is immediately clear that in the sense of the
term 'human being' which is required to make the first premise acceptable,
the second premise is false. The embryo, especially the early embryo, is
obviously not a being with the mental qualities that generally distinguish
members of our species from members of other species. The early embryo
has no brain, no nervous system. It is reasonable to assume that, so far as
its mental life goes, it has no more awareness than a lettuce.

It is still true that the human embryo is a member of the species *Homo
sapiens.* That is why it is difficult to deny that the human embryo is a human
being. But we can now see that this is not the sense of 'human being' needed
to make the standard argument work. A valid argument cannot equivocate
on the meanings of its central terms. If the first premise is true when 'human'
means 'a being with certain mental qualities' and the second premise is true
when 'human' means 'member of the species *Homo sapiens*', the argument
is based on a slide between the two meanings, and is invalid.

Can the argument be rescued? It obviously can't be rescued by claiming
that the embryo is a being with the requisite mental qualities. That might
be arguable only for some later stage of the development of the fetus. If the
second premise cannot be reconciled with the first, then, can the first perhaps
be defended in a form which makes it compatible with the second? Can it
be argued that human beings have a right to life, not because of any moral

qualities they may possess, but because they—unlike pigs, cows, dogs or lettuces—are members of the species *Homo sapiens*?

This is a desperate move. Those who make it find themselves having to defend the claim that species membership is in itself morally relevant to the wrongness of killing a being. But why should that be so? If we are considering whether it is wrong to destroy something, surely we must look at its actual characteristics, not just the species to which it belongs. If visitors from other planets turn out to be sensitive, thinking, planning beings, who form deep and lasting relationships just like we do, would it be acceptable to kill them simply because they are not members of our species? What if we substituted 'race' for 'species' in the question? If we reject the claim that membership of a particular race is in itself morally relevant to the wrongness of killing a being, it is not easy to see how we could accept the same claim when based on species membership. The fact that other races, like our own, can feel, think and plan for the future is not relevant to this question, for we are considering membership in a particular group—whether race or species—as the sole basis for determining the wrongness of killing members of one group or another. It seems clear that neither race nor species can, in itself, provide any justifiable basis for such a distinction.

So the standard argument fails. It does so not because the embryo is not a human being, but because the sense in which the embryo is a human being is not the sense in which we should accept that every human being has a right to life.

III

We have now seen the inadequacies of arguing that the human zygote or early embryo is a distinct human individual, and that destructive embryo experimentation is wrong because the zygote or embryo is a member of the human species. But there are other reasons why one might consider embryo experimentation wrong. Since the early embryo, devoid of a nervous system or a brain, can neither experience pain or pleasure, nor any of the things occurring in the world, the most important thing about it is that it is a potential baby or person, a person just like us. In other words, when we destroy an early human embryo in research, a potential baby or person will now not exist.

Why is this fact morally significant? One plausible answer is provided by R.M. Hare when he appeals, in a well-known article on abortion, to a type of formal argument, captured in the ancient Christian and pre-Christian Golden Rule, that has been the basis of almost all theories of moral reasoning: that we should do to others as we wish them to do to us.[6] In other words, given that we are glad that nobody destroyed us when we were embryos, we should, other things being equal, not destroy an embryo that would have a life like us.

It might seem that the Golden Rule applied to embryo experimentation would impose on us an extremely conservative position, for it would seem to rule out the destruction of all but the most seriously abnormal zygote or embryo. But before we too readily embrace that conclusion, we should also note something already pointed out by Hare in the abortion context: when you are glad that you exist, you do not confine this gladness to gladness that you were not aborted when an embryo or fetus. Rather, you are also glad that you were brought into existence in the first place—that your parents had intercourse without contraception when they did.[7]

Let us apply this sort of thinking to our present context—that of the IVF embryo—and assume that, at some time between the beginning of the process of fertilization and the formation of the primitive streak, the existence of an identifiable individual began. Let us also assume that you developed from that individual. Now, it will immediately become apparent that regardless of what event marked the beginning of your life as an identifiable entity, none of the other events that preceded it were any less important for your present existence. Just as you would not have existed had a scientist performed a destructive experiment on the embryo from which you developed when it was 14 days old, so you would not have existed had she or he performed it on the zygote when it was one day old. Similarly, you would not have existed had your mother's egg with your father's sperm already inside it been destroyed just before syngamy had occurred, or just after that event. Nor, we should hasten to add, would you have existed had the scientist, instead of using one particular egg, used another egg or had a different sperm fertilized the egg, and so on.

The upshot is that the marker event for the beginning of a human individual, on whose identification so much time and energy is being expended, is of no importance so far as the existence of a particular person is concerned. If it is the existence of a particular person that is relevant—a Tom, a Dick, or a Harry—who would treasure his life in much the same way as we do, then it does not matter whether his existence was thwarted before or after fertilization, or the formation of the primitive streak, had occurred.

We should also note that there are numerous ways in which the existence of particular individuals can be thwarted. A totipotent IVF embryo, for example, is not, as we saw, one particular individual, but rather an entity with the potential to become one or more different individuals, because each cell is a distinct entity with the potential for further development. What are we doing, then, when we refrain from separating the cells, leaving all four or eight of them together? And what are we doing when we extract and destroy a single cell for gene-typing of the embryo? We believe we should, in consistency, say that we are depriving a number of human individuals of their chance of existence.

This is one important point. The other important point is this: our reproductive choices almost invariably constitute an explicit or implicit choice between different individuals. We said a moment ago that had the scientist who assisted your imaginary IVF conception used a different sperm or a different egg, you would not have existed, and the same thing applies, of course, to other scenarios in natural reproduction as well. But—and this is the morally important point—seeing that your parents wanted to raise a child of their own, it is likely that another child would have been born. While this person would not have been you, it would have been a person just like you in the morally relevant sense that she or he would now, presumably, be just as glad to exist as you are glad to exist.

But if our reproductive choices typically constitute a choice between different individuals, then the destruction of early human embryos—particularly if it makes possible improved IVF techniques and, therefore, the existence of IVF children who would not otherwise have existed—is no harder to justify than many of our other reproductive choices: for example, when and with whom, to have intercourse, without contraception, to have the two or three children we are going to have.

IV

We have now seen that some of the most common objections to destructive experimentation on early human embryos are seriously flawed. But when, in its development from zygote to baby, does the embryo acquire any rights or interests? We believe the minimal characteristic needed to give the embryo a claim to consideration is sentience, or the capacity to feel pleasure or pain. Until that point is reached, the embryo does not have any interests and, like other non-sentient organisms (a human egg, for example), cannot be harmed—in a morally relevant sense—by anything we do. We can, of course, damage the embryo in such a way as to cause harm to the sentient being it will become, if it lives, but if it never becomes a sentient being, the embryo has not been harmed, because its total lack of awareness means that it never has had any interests at all.

The fact that the early embryo has no interests is also relevant to a distinction embodied in the *Infertility (Medical Procedures) Act 1984* (Vic.) between 'spare' embryos left over from infertility treatments (which may be used in experimentation), and the creation of embryos especially for research (which is prohibited).[8] The report of the Waller Committee, on which the legislation is based, speaks of such a creation as using a human being as a means rather than as an end.[9] This is a principle of Kantian ethics that makes some sense when applied to rational, autonomous beings—or perhaps even, though more controversially, when applied to sentient beings who, though not rational or autonomous, may have ends of their own. There is no basis at all, however, for applying it to a totally non-sentient embryo, which can have no ends of its own.

Finally, we point to a curious consequence of restrictive legislation on embryo research. In sharp contrast to the human embryo at this early stage of its existence, non-human animals such as primates, dogs, rabbits, guinea pigs, rats and mice clearly can feel pain, and thus often are harmed by what is done to them in the course of scientific research. We have already suggested that the species of a being is not, in itself, relevant to its ethical status. Why, then, is it considered acceptable to poison conscious rabbits in order to test the safety of drugs and household chemicals, but not considered acceptable to carry out tests on totally non-sentient human embryos? It is only when an embryo reaches the stage at which it may be capable of feeling pain that we need to control the experimentation which can be done with it. At this point the embryo ranks, morally, with those non-human animals we have mentioned. These animals have often been unjustifiably made to suffer in scientific research. We should have stringent controls over research to ensure that this cannot happen to embryos, just as we should have stringent controls to ensure that it cannot happen to animals.

　　At what point, then, does the embryo develop a capacity to feel pain? Though we are not experts in this field, from our reading of the literature, we would say that it cannot possibly be earlier than six weeks, and it may well be as late as 18 or 20 weeks.[10] While we think we should err on the side of caution, it seems to us that the 14-day limit suggested by both the Waller and Warnock committees is too conservative.[11] There is no doubt that the embryo is not sentient for some time after this date. Even if we were to be very, very cautious in erring on the safe side, a 28-day limit would provide sufficient protection against the possibility of an embryo suffering during experimentation.

Notes

1 Senate Select Committee on the Human Embryo Experimentation Bill 1985, *Human Embryo Experimentation in Australia* (Senator Michael Tate, chairman), (AGPS, Canberra, 1986), p. 13.

2 Ibid.

3 Congregation for the Doctrine of the Faith, *Instruction on Respect for Human Life in its Origin and on the Dignity of Procreation—Replies to Certain Questions of the Day* (Vatican City, 1987).

4 Section I of this chapter was inspired by Michael Coughlan's: "'From the Moment of Conception . . .": The Vatican Instruction on Artificial Procreation Techniques', *Bioethics* II, 3 (July 1988).

5 For a more detailed discussion of this issue, see Norman Ford, *When did I begin?* (Cambridge University Press, Cambridge, 1988).

6 Hare, R.M., 'Abortion and the golden rule', *Philosophy and Public Affairs* 4: 3 (1975), 201–22.

7 Ibid, p. 212.

8 *Infertility (Medical Procedures) Act 1984* (Vic.).

9 Committee to Consider the Social, Ethical and Legal Issues Arising from In Vitro Fertilization, *Report on the Disposition of Embryos Produced by In Vitro Fertilization* (Prof. Louis Waller, chairman), (Victorian Government Printer, Melbourne, 1984), para. 3.27.

10 For an expert opinion on when a fetus may begin to be capable of feeling pain, see the report of the British Government's Advisory Group on Fetal Research, *The Use of Foetuses and Foetal Material for Research* (Sir John Peel, chairman), (HMSO, London, 1972). A clear summary of some relevant scientific evidence, with further references, can be found in M. Tooley, *Abortion and Infanticide* (Clarendon Press, Oxford, 1983), pp. 347–407.

11 Waller, op. cit., para. 3.29; and *Report of the Committee of Inquiry into Human Fertilization and Embryology* (Mary Warnock, chair), (HMSO, London, 1984), pars. 11.19–11.22.

8 | IVF technology and the argument from potential

PETER SINGER AND KAREN DAWSON

I

In many respects the current debate about embryo experimentation resembles the older debate about abortion. Although one central argument for abortion—the claim that a woman has the right to control her own body—is not directly applicable in the newer context, the argument against embryo experimentation remains essentially the same as the argument against abortion. This argument has two forms, one relying on the claim that from the moment of fertilization the embryo is entitled to protection because it is a human being, and the other asserting instead that the embryo is entitled to protection because from the moment of fertilization it is a potential human being.[1]

The first form of this argument is considered in Chapter 5; our focus is on the argument from potential. Those who use this argument against embryo experimentation frequently describe the potential of the early *in vitro* embryo in terms which are identical with the terms used to describe the potential of the early embryo inside the female body in the context of the abortion debate. For example, Teresa Iglesias, writing on *in vitro* fertilization, says: 'We know that a new human individual organism with the internal potential to develop into an adult, given nurture, comes into existence as a result of the process of fertilization at conception.'[2]

But can the familiar claims about the potential of the embryo in the uterus be applied to the embryo in culture in the laboratory? Or does the new technology lead to an embryo with a different potential from that of embryos made in the old way? Asking this question leads us to probe the meaning of the term 'potential'. This probing will raise doubts about whether it is meaningful to talk of the potential of an entity, independently of the context in which that entity exists, and independently of the probability of that entity developing in a specific way. In particular, we will argue that while the notion of potential may be relatively clear in the context of a naturally occurring process such as the development of an embryo inside a female body, this notion becomes far more problematic when it is extended to a laboratory situation, in which everything depends on our knowledge and skills, and on what we decide to do. This line of argument will lead us to the conclusion that there is no coherent notion of potential which allows the argument from potential to be applied to embryos in laboratories in the way in which those who invoke the argument are seeking to apply it.

We begin by considering how recent developments in reproductive technology force us to revise some previously universal truths about embryos. Before Robert Edwards began the research which was to lead to the IVF procedure, no one had observed a viable human embryo prior to the stage at which it implants in the wall of the uterus. As long as pre-implantation embryos existed only inside the woman's body, there was no way of observing them during that period. The very existence of the embryo could not be established until after implantation.

Under these circumstances, once the existence of an embryo was known, that embryo had a good chance of becoming a person, unless its development was deliberately interrupted. The probability of such an embryo becoming a person was, therefore, very much greater than the probability of an egg in a fertile woman uniting with sperm from that woman's partner and leading to a child. It was also considerably greater than the chances of an as yet unimplanted embryo becoming a child.

There was also, in those pre-IVF days, a further important fact, independent of the difference in probability, between any embryo, whether implanted or not, and the egg and sperm. Whereas the embryo inside the female body has some definite chance (we shall consider later how great a chance) of developing into a child unless a deliberate human act interrupts its growth, the egg and sperm can only develop into a child if there is a deliberate human act. So in the one case, all that is needed for the embryo to have a prospect of realizing its potential is for those involved to refrain from stopping it; in the other case, they have to carry out a positive act. The development of the embryo inside the female body can, therefore, be seen as a mere unfolding of a potential that is inherent in it. The development of the separated egg and sperm is more difficult to regard in this way, because no further development will take place unless the couple have sexual intercourse or use artificial insemination. (Admittedly, this is an oversimplification, for it takes no account of the positive acts involved in childbirth; but it is close enough for our purposes.)

Now consider what has happened as a result of the success of IVF. The procedure involves removing one or more eggs from a woman's ovary, placing them in culture medium in a glass dish, and then adding sperm to the culture. In the more proficient laboratories, this leads to fertilization in about 80% of the eggs thus treated. The embryo can then be kept in culture for 2–3 days, while it grows and divides into two, four and then eight cells. At about this stage, if the embryo is to have any prospect of developing into a child, it must be transferred to a woman's uterus. Although the transfer itself is a simple procedure, it is after the transfer that things are most likely to go wrong: for reasons which are not fully understood, with even the most successful IVF teams, the probability of a given embryo which has been transferred to the uterus actually implanting there, and leading to a continuing pregnancy, is always less than 20%, and generally no more than 10%.[3] (Figures quoted for pregnancies per transfer procedure may be higher, but this is because it is common to transfer more than one embryo; for our purposes the important figure is the probability of any given embryo resulting in a child.) We should also note that if the embryo is allowed to continue to grow in culture much beyond the eight-cell stage, it is less likely to implant when transferred. Embryos can be grown in the laboratory to the later blastocyst stage, when the cells are arranged as a hollow sphere and those which will form the embryo proper have become distinct from those which will form the extra-embryonic membranes, that is, the chorion and the amnion. The blastocyst may then develop further, to the point at which it consists of hundreds of cells. No pregnancies, however, have resulted from embryos transferred at so late a stage of development. Nor, as yet, is there any prospect of keeping embryos alive and developing *in vitro* until they become viable infants. So while Edwards has reported keeping an embryo alive in culture for nine days,[4] with our present state of knowledge, such an embryo has zero probability of becoming a person.

In summary, then, before the advent of IVF, in every instance in which we knew of the existence of a normal human embryo, it would have been true to say of that embryo that, unless it was deliberately interfered with, it would most likely develop into a person. The process of IVF, however, leads to the creation of embryos which cannot develop into a person unless there is some deliberate human act (the transfer to the uterus) and which even then, in the best of circumstances, will most likely not develop into a person.

The upshot of all this is that IVF has reduced the difference between what can be said about the embryo, and what can be said about the egg and sperm, considered jointly. Before IVF, any normal human embryo known to us had a far greater chance of becoming a child than any egg plus sperm prior to fertilization taking place. But with IVF, there is a much more modest difference in the probability of a child resulting from a two-cell embryo in a glass dish, and the probability of a child resulting from an egg and some sperm in a glass dish. To be specific, if we assume that the laboratory's fertilization rate is 80% and its rate of pregnancy per embryo transferred is 10%, then the probability of a child resulting from a given embryo is 10%, and the probability of a child resulting from an egg which has been placed in a culture medium to which sperm has been added is 8%.

II

It has occasionally been suggested that there is no difference between the potential of the embryo, on the one hand, and of the egg and sperm when still separate, but considered jointly, on the other hand.[5] But there has been little analysis of the notion of potential in the context of the *in vitro* embryo, and the suggestions made have not succeeded in dispelling the intuitive idea that there is a major difference between the potential of the embryo and the potential of the pair of gametes. To provide this analysis, we must now ask: what do we mean when we refer to the embryo as a potential person?

An obvious place to begin our search for the meaning of this claim is with the dictionary definition of the word 'potential'. The *Oxford English Dictionary* offers several meanings of the term, of which the following seems to be the most relevant to our present concerns: '*Potential . . .* Possible as opposed to actual; existing *in posse* or in a latent or undeveloped state, capable of coming into being or action; latent.' Following the dictionary definition, it would seem that as a minimum we must be saying that it is possible for the embryo to become a person. Possibility is a necessary condition for potentiality. (Whether it is also a sufficient condition is not something we need to consider here.) But what sort of possibility is this?

Philosophers commonly distinguish between logical possibility and physical possibility. It is logically possible, but physically impossible, for the authors of this paper to jump over the Empire State Building. It is both logically and physically possible for us to jump over a brick. It is not logically possible for anyone to be a biological parent without having any children.

Since something is logically impossible only if its assertion involves a contradiction, there is nothing logically impossible about a human blastocyst in a laboratory developing into a person. But then, there is nothing logically impossible about a human egg developing into a person either—parthenogenesis happens often in some species, and certainly no logical contradiction is involved in imagining it happening in our species. So those who claim that the human embryo is a potential person, whereas the human egg is not, cannot appeal to the mere logical possibility of the embryo becoming a person.

The sense of 'possibility' that lies behind these claims that the embryo, but not the egg, is a potential person must, therefore, be real, physical possibility, and not logical possibility. Now, however, we must further refine the relevant sense of physical possibility. Does it refer to what is physically possible, given the present state of our knowledge and technology? In that case, the eight-cell embryo in the laboratory may be a potential person; but a late-stage blastocyst in the laboratory, consisting of hundreds of cells, cannot be a potential person; we know that if we attempt to transfer such a blastocyst, it will simply be discharged from the uterus without implanting. This yields the result that two blastocysts, to all appearances identical in their internal properties, may have entirely different potentials. One, because it has resulted from natural intercourse and has implanted in the uterus, can be a potential person, while the other one cannot be because it is in a laboratory culture.

Such a result is counterintuitive, for it means that while the eight-cell embryo in the laboratory is a potential human being, the embryo loses that status simply by continuing to develop in the laboratory. But perhaps we could come to accept such a view. There are analogous situations in which we would also say that a being has lost the potential it once had. Imagine, for instance, a doctor monitoring a risky pregnancy. The doctor might observe a healthy fetus at one stage during the pregnancy, and say: 'Yes, we have a potential person there.' Gradually, however, the condition of the fetus may deteriorate to the point at which it is evident that the fetus will die before reaching the point at which a caesarean delivery could offer any hope of producing a viable infant. At that point the doctor might say that the potential for personhood has been lost.

This account of potentiality may appear to confuse potential with probability. So far, however, we have been doing no more than exploring a minimum necessary condition, suggested by the dictionary definition, of X having the potential to become Y. That minimum condition is that it be possible for X to become Y. Once we accept that it is a present physical possibility, and not logical possibility that is meant here, we cannot disregard the differences between the eight-cell embryo in the laboratory and those blastocysts which consist of hundreds of cells. These differences do mean that, given our present state of knowledge and technology, it is possible for the former to become a person, but quite impossible, in the relevant physical sense, for the blastocyst just described to become a person. If physical possibility in our present state of knowledge and technology is a necessary condition for potentiality, it follows that the blastocyst in the laboratory is not a potential person.

In view of the implications of this view, it might be said that the relevant sense of 'physically possible' should not refer to the present state of our knowledge and technology. If we should one day discover how to induce late blastocysts to implant, or if we should perfect laboratory development to such an extent that embryos can develop into infants without ever being transferred to a woman—a process known as ectogenesis—then late blastocysts in laboratories will be able to become people. Perhaps this is all the 'possibility' that is needed for an embryo to be a potential person.

This may indeed be the sense of 'possibility' which lies behind a proper attribution of potential; but it cannot help those who wish to distinguish the potential of the embryo from that of the egg alone. For if it is true that we may one day discover how to induce late blastocysts to implant, it is also true that we may one day discover how to induce parthenogenetic development—spontaneous development in the absence of any sperm—in the human egg. (Scientists putting human eggs in culture media for IVF have reported seeing, on rare occasions, the beginnings of parthenogenetic development.[6]) So the same sense of 'possibility' which would allow the late embryo to be a potential person would also allow every human egg to be a potential person.

At one stage in the development of reproductive technology—roughly from 1983 until 1985—it might have been argued that the late blastocyst had a genuine possibility of becoming a person in a way that the egg did not. During that brief period, human embryos had successfully been preserved by freezing, in a manner which made it possible for them to continue normal development after thawing. There was, however, then no known way of freezing human eggs which did not cause them damage so severe as to make continued development impossible. A blastocyst could, therefore, have been frozen to await discovery either of a technique for implanting it successfully in a uterus, or of the means of developing it to viability in an artificial womb. A human egg could not have been frozen to await the development of a means of inducing parthenogenesis. Since 1985, however, it has been possible to freeze eggs as well as embryos. So if the combination of freezing and future discoveries means that a laboratory blastocyst is now a potential person, the same combination must now mean that an unfertilized egg in a laboratory is also now a potential person.

Unravelling the notion of potential is leading us in an unexpected direction, and one which will not be welcomed by those who oppose experimentation on human embryos, while permitting experimentation on human eggs.[7] The problem, however, is not with the analysis we have proposed, but with the attempt to develop a notion of potential which supports the idea that there is a sharp distinction between the potential of the embryo, and that of either the separate egg, or of the egg and sperm when separate but considered jointly. Is there any way in which the notion can be restored to something more suitable to those purposes?

One writer, Warren Quinn, has discussed parthenogenesis.[8] He suggests that in this situation the environmental agent producing parthenogenetic development can be treated as a pre-fertilization entity that is incorporated into the 'zygote' at the onset of development. In this way, he seeks to preserve the view that even if parthenogenesis occurs, the egg alone is not a potential person; it only becomes a potential person once parthenogenetic development has been triggered. But Quinn's suggestion does not succeed in marking a distinction between the egg and the embryo. For the embryo also needs a specific environment if it is to develop; and if the particular environment which leads to parthenogenesis is allowed to count as an entity for the purposes of denying potential to the egg on its own, outside that environment, then the particular environment which leads to development of the embryo must also be allowed to count as an entity and would imply that we should deny potential to the embryo on its own, outside that environment.

One might try to defend Quinn's analysis by claiming that the embryo has an inherent potential to develop into a person; the egg, on the other hand, needs an external trigger if it is to develop. At first glance, this appears promising; but on closer scrutiny the promise evaporates. Both the egg and the embryo have an internal genetic code which can, in the right environment, lead to the development of a human being. True, in the case of the embryo, the 46 chromosomes are already present, whereas in the case of the egg, the 23 chromosomes which are present will need to duplicate themselves to form the 46 chromosomes necessary for further development. But in neither case does any additional genetic information have to be supplied from any external source. In both cases, on the other hand, a great deal else does have to come from outside. In the case of the embryo in the uterus, this includes all the nutrients needed for growth; and of course in the case of the embryo in the laboratory, it also includes skilled human intervention to transfer the embryo to a uterus. In the case of the egg, skilled human intervention would also be required to stimulate intentional parthenogenetic development. The difference seems to be one of degree rather than of kind.

It might be said that the induction of parthenogenesis marks a more radical change than that caused by the provision of nutrients, because it marks the beginning of a new individual, and in this respect parthenogenesis and fertilization are alike, while the subsequent stages of growth and development have a different, and lesser, significance. But why should we regard the egg after parthenogenesis as a different individual from the egg before parthenogenesis? The following reason can be offered: before either fertilization or parthenogenesis, the egg could develop into any number of different people (because it could be fertilized by any one of a number of different sperm, or could develop parthenogenetically). After fertilization or parthenogenesis, the developing embryo can become only one person. (This is not strictly true, because of the possibility of twinning, but the contrast remains, between an indefinite range of possibilities, and a very limited range of possibilities.)

In our view the fact that the embryo, but not the egg, has a uniquely determined potential does not suffice to show that the embryo is a different individual from the egg, or that it, but not the egg, is a potential person. Consider the analogy of a block of marble, rough-hewn from the quarry. In the hands of Michelangelo, it is a potential *David*, or a *Moses*, or a *Pietà*. Later, when the sculptor has chiselled it into the rough outline of a standing, youthful figure, it can only become a *David*. Certainly, by working the marble in this way, Michelangelo has taken its development a stage further. The stage is significant because now the marble has the potential to become only one kind of sculpture (though of course there is still scope for great variation in many important details). Yet the marble is continuous in space and time with the original block. It is not a different piece of marble. That original block, we can now see, had the potential to be a *David* all along, and the fact that at an earlier stage it could also have become something else does not count against the claim that, even then, it was a potential *David*. Similarly, fertilization or parthenogenesis takes the development of the egg a stage further, but the potential of the egg is retained. The resulting embryo now has the potential to become only one kind of person (though here too there is still scope for great variation in many important details[9]). Yet the egg had the potential to become this person all along, just as it had the potential to become, in different circumstances, any one of a wide range of other people. Potentiality is one thing; uniqueness is something quite different.

Although we have used the possibility of parthenogenetic development as a means of illustrating some of the problems of attempts to separate the potential of the late blastocyst from the potential of the human egg, our general analysis of the notion of potential does not rely on this. We could equally well have returned to the simpler case of the egg and sperm, together in their culture medium, but prior to the occurrence of fertilization. For all the senses of 'possibility' that we have considered, it is no less possible for the egg and sperm in the laboratory to develop into a person than it is for the laboratory embryo, also in its culture medium, to develop into a person. One could even say the same about the egg alone, treating the presence of sperm as part of the environment necessary for further development, just as the presence of nutrients is necessary for the further development of the embryo.

A more promising attempt to distinguish the potential of the embryo from the potential of the egg and sperm in their culture medium is openly to acknowledge a link between potential and probability, by relating potential not to the bare possibility of the embryo, or whatever other entity we are considering, becoming a person, but rather to the probability of this happening. This has the inevitable result that potential ceases to be an all-or-nothing matter, and becomes a matter of degree. Traditional defenders of the right to life of the embryo have been reluctant to introduce degrees of potential into the debate, because once the notion is accepted, it seems undeniable that the early embryo is less of a potential person than the later embryo or the fetus. This could easily be understood as leading to the conclusion that the prohibition against destroying the early embryo is less stringent than the prohibition against destroying the later embryo or fetus. Nevertheless, some defenders of the argument from potential have invoked probability and degrees of potential. Among those who have spoken most openly of probability have been the Roman Catholic theologian John Noonan and the philosopher Werner Pluhar. As Noonan puts it:

> *As life itself is a matter of probabilities, as most moral reasoning is an estimate of probabilities, so it seems in accord with the structure of reality*

and the nature of moral thought to found a moral judgment on the change in probabilities at conception . . . Would the argument be different if only one out of ten children conceived came to term? Of course this argument would be different. This argument is an appeal to probabilities that actually exist, not to any and all states of affairs which may be imagined . . . If a spermatozoon is destroyed, one destroys a being which had a chance of far less than 1 in 200 million of developing into a reasoning being, possessed of the genetic code, a heart and other organs, and capable of pain. If a fetus is destroyed, one destroys a being already possessed of the genetic code, organs and sensitivity to pain, and one which had an 80 per cent chance of developing further into a baby outside the womb who, in time, would reason. [10]

Pluhar is almost as explicit:

. . . if we allow a mere potential for simple consciousness to give rise to a prima facie right to life, then it seems that we must accord a similar right to the staggering number of gamete pairs that likewise have some such potential . . . Clearly, however, the gamete pair's potential is vastly lower than that of the insentient fetus: even given absence of interference plus at most a modest amount of assistance, the probability of a given gamete pair's producing the individual that it has some potential to produce is so vanishingly small as to be totally negligible in practice. [11]

If, following Noonan and Pluhar, we take the probability of an embryo becoming a reasoning being (or, in Pluhar's case, becoming sentient) as relevant to the potential of the embryo to become a person, it must follow that the potential of the laboratory embryo currently diminishes after the eight-cell stage, when the probability of the transferred embryo resulting in a pregnancy begins to decline; and by the late blastocyst stage, on this view, the laboratory embryo has no potential at all. This may well be an implication which opponents of embryo experimentation are happy to accept; they may say that this loss of the potential to become a person is one reason why it is wrong to keep human embryos alive in laboratories, or perhaps even why it is wrong to create them *in vitro* at all.

Accepting that there are degrees of potential associated with probability does, however, have other consequences which are less likely to be congenial to opponents of embryo experimentation. For on this view, contrary to what Noonan and Pluhar claim, the distinction between the potential of the embryo in culture, and the potential of the gametes in the laboratory before fertilization, becomes a difference of degree, and not a marked difference at that. Fertilization is, as we have seen, one of the relatively reliable steps in the *in vitro* fertilization procedure, with success rates commonly around 80%. Thus if we are to base degrees of potential on the probability of a person ultimately resulting from an embryo, we could not treat as crucially signifi-cant the line between the stage at which we have a set of gametes and the stage at which we have an embryo. At least so far as potential is concerned, a much more significant division would be between the stage at which we have an embryo in the laboratory, and the stage at which we have an embryo implanted in the uterus. In so far as the argument from potential is important to the morality of experimenting on, or disposing of, an entity, we could not support the prohibition of experimenting on, or disposing of, embryos, while remaining unconcerned about how eggs and sperm are treated.

There are two possible replies to this argument. The first claims that to speak of the potential of the egg and sperm, while they are still separate, is nonsense, because they are two discrete entities, and hence cannot have a single potential. The second reply is Noonan's; it asserts that the distinction between embryo and gametes does mark a sharp distinction in probability, because the probability of an embryo becoming a child is very great, whereas the probability of any one sperm participating in fertilization is one in 200 000 000. We will consider these replies in turn.

The first reply fails, because there is no reason why an entity with potential must consist of a single object, rather than of two or more discrete objects. There is, for instance, nothing problematical about the statement (made, let us assume, shortly before the battle of El Alamein): 'Montgomery's army has the potential to defeat Rommel's army.'[12] Yet Montgomery's army consisted of thousands of discrete individuals, spread over many miles of desert. We can even speak of the potential of entities which are spread across the entire planet—as Noah might have spoken of the potential of the raindrops falling all over the world to cause a great flood. So why should there be any problem about speaking of the potential of a set of gametes in a glass dish?

Noonan's reply faces several problems. One has been raised by Mark Strasser.[13] Why, Strasser asks, does Noonan focus on the probability of a single sperm participating in fertilization, and not on the probability of fertilization by any one of the sperm? This would, of course, provide a very different answer: in the case of a normally fertile woman having sexual intercourse without contraception during that part of her cycle when she is most likely to be fertile, the probability of fertilization taking place and resulting in a child is not so greatly different from the probability of the newly fertilized egg resulting in a child—certainly not different by the orders of magnitude Noonan suggests. Similarly, Noonan does not discuss the probability of the egg rather than the sperm participating in fertilization. This would also give a very different result.

Even if Noonan can provide an answer to Strasser's objection, his position has, like other claims about the potential of embryos, become much more difficult to maintain in the light of new knowledge and new developments in reproductive technology. The initial difficulty is that Noonan's figures for embryo survival even in the uterus are no longer regarded as accurate.

At the time Noonan wrote, the estimate of pregnancy loss was based on clinically recognized or stable, continuing pregnancies. These pregnancies are about 6–8 weeks after fertilization—embryonic heart-beat is detectable, menses has ceased and enzyme assays will give reliable results indicating pregnancy. Currently such pregnancies are associated with a 15% loss through spontaneous abortion.[14] Even though the total pregnancy wastage rate remains largely unknown, recent technical advances allowing earlier recognition of pregnancy suggest that this figure is an underestimate of loss and an oversimplification of the real situation.[15] Estimates of the natural wastage at various stages of pregnancy can now be taken into account and this provides startlingly different figures from those supplied by Noonan. If pregnancy is diagnosed before implantation (within 14 days of fertilization) the estimated chance of a birth resulting is 25–30%.[16] Post-implantation this chance increases initially to 46–60%,[17] and it is not until six weeks gestation that the chance of birth occurring increases to 85–90%.[18]

Noonan claimed that his argument is '. . . an appeal to probabilities that actually exist, not to any and all states of affairs which may be imagined'. We have now seen that the real probabilities are very different from what Noonan believed them to be. Once we substitute the real probabilities of embryos, at various stages of their existence, becoming persons, Noonan's argument no longer supports the moment of fertilization as the time at which the embryo gains a significantly different moral status. Indeed, if we were to require an 80% probability of further development into a baby—the figure used in the passage from Noonan quoted above—we would have to wait until about six weeks after fertilization before the embryo would have the significance Noonan wants to claim for it. If, on the other hand, we simply look for the moment at which the chance of birth resulting first becomes close to or better than 50%, that time would seem to be around implantation.

To cope with the development of IVF, some readjustment of the parts of this argument pertinent to gametes is also necessary. Most importantly, the figures for embryo survival are very different when we consider the laboratory embryo rather than the embryo implanted in a uterus; an embryo survival rate of 10% would be relatively optimistic, even in a proficient laboratory. In addition, Noonan estimates the probability of any one sperm participating in fertilization as one in 200000000 which is based on the number of sperm in a male ejaculate. However, in IVF only about 50000 sperm are used to fertilize an egg which, comparatively, greatly increases the chances for any one sperm fertilizing the egg.[19]

Perhaps Noonan's claim of a sharp difference between the embryo and the sperm, based on the probability of proceeding to the next stage of development, could survive these changes in the figures. The relevant figure for the embryo *in vitro* is now 1 in 10, and for the sperm participating in *in vitro* fertilization, about 1 in 50000. This is still a very marked difference. The difference would virtually disappear, however, if we focused on the egg rather than the sperm; or if, as Strasser suggests, we consider the prospects of a birth resulting not from a given sperm, but from one or other of the sperm in the seminal fluid.

In any case, there is still one more difficulty for the argument. Scientists are at present trying out a new means of overcoming male infertility caused by a low sperm count or sperm which is insufficiently motile. The egg will be removed and cultured as in the normal *in vitro* procedure; but instead of adding a drop of seminal fluid containing about 50000 sperm, a single sperm will be microinjected under the outer membrane of the egg. This procedure has already been carried out with human gametes, although no live births have yet resulted.[20] Using the technique to overcome male infertility may encounter problems; but assuming that it is successful, the unique genetic blueprint of the individual-to-be will be determined before fertilization; it will, to be precise, be determined at the moment when the single sperm has been selected for microinjection. So if we compare the probability of the embryo becoming a person with the probability of the egg, together with the single sperm about to be microinjected into the egg, becoming a person, we will be unable to find any sharp distinction between the two. They cannot even be distinguished in terms of the genetic blueprint having been determined in one case but not the other.

III

Where does this leave the notion of potential, as applied to an embryo in a laboratory? We noticed earlier that whereas the embryo inside the female body has some definite chance of developing into a child unless a deliberate human act interrupts its growth, the egg and sperm can only develop into a child if there is a deliberate human act. In this respect the embryo in the laboratory is like the egg and sperm, and not like the embryo in the human body. This is of fundamental importance for the notion of potential, because lurking in the background of discussions of the embryo's potential is the idea that there is a 'natural' course of events, governed by the 'inherent' potential of the embryo. We have seen, however, that this notion of 'natural' development, not requiring the assistance of a deliberate human act, has no application to the IVF embryo. Hence those who wish to use the potential of the IVF embryo as a ground for protecting it cannot appeal to this notion of natural development; and for this reason, they find themselves in difficulty in explaining why the embryo in the laboratory has a potential so different either from that of the egg alone, or of the egg and sperm considered jointly.

Unless a woman agrees to have an embryo transferred to her uterus, and someone else agrees to perform the transfer, that embryo has no future. This suggests that it may be more enlightening to focus on who should make such decisions, and on what grounds, than on the issue of the potential of the laboratory embryo. The question about potential may simply be insoluble when considered in abstraction from human decisions.[21]

We have examined a range of possible arguments which invoke the potential of the embryo as a reason for according it a special moral status, different from that of the egg, or of the egg and sperm when separate, but considered jointly. We have looked at whether potential can be based on logical possibility, on physical possibility relative to our present state of knowledge, on physical possibility independent of our present state of knowledge, or on probability; and we have also looked at the claim that the embryo has an inherent potential which the egg and sperm lack, because only the embryo has the ability to develop without further deliberate human action. We have not offered an opinion on whether these arguments succeed in establishing that, in the normal reproductive situation, the embryo has a potential different from that of the egg and sperm. We have, however, given reasons why, even if these arguments are successful in the normal situation, they cannot validly be applied to *in vitro* embryos and eggs and sperm. The new reproductive technology makes it necessary to think again about how our established views about the potential of the human embryo should be applied to the embryo in a laboratory.

Notes

1 For examples of the popular arguments against embryo experimentation, see William Walters and Peter Singer (eds), *Test-tube Babies* (Oxford University Press, Melbourne, 1982), ch. 4.

2 Iglesias, Teresa, '*In vitro* fertilization: the major issues', *Journal of Medical Ethics* 10 (1984), 36.

3 For these figures, see Ian Johnston, 'IVF: the Australian experience', a paper presented at the Royal College of Gynaecologists and Obstetricians Study Group on AID and IVF, November 1984, and reprinted in Hansard, Senate Select Committee on the Human Embryo Experimentation Bill 1985, report of hearings of 26 February 1986 (AGPS, Canberra, 1986), pp. 560–87.

4 Edwards, Robert and Steptoe, Patrick, *A Matter of Life* (Sphere, London, 1981), p. 146.

5 For example, Helga Kuhse and Peter Singer, 'The moral status of the embryo' in Walters and Singer, op. cit. (see note 1 above), pp. 57–63; John Robertson, 'Extracorporeal embryos and the abortion debate', *Journal of Contemporary Health Law and Policy* 2 (1986), 63.

6 Edwards, R., and Trounson, A., 'Discussion' in R. Edwards and J. Purdy (eds), *Human Conception in Vitro* (Academic Press, London, 1982), pp. 219–33.

7 As was recommended by the Victorian Government's Committee to Consider the Social, Ethical and Legal Issues Arising from *In Vitro* Fertilization, chaired by Professor Louis Waller. See the committee's *Report on the Disposition of Embryos*

Produced by In Vitro Fertilization (Victorian Government Printer, Melbourne, 1984). The subsequent Victorian legislation, the *Infertility (Medical Procedures) Act 1984*, Section 6, incorporates these recommendations by tightly restricting embryo experimentation while explicitly exempting experimentation on human ova.

8 Quinn, Warren, 'Abortion: identity and loss', *Philosophy and Public Affairs* 13 (1984), 28.

9 On the range still possible after fertilization, see Karen Dawson, 'Fertilization and moral status: a scientific perspective', Chapter 5 of this book.

10 Noonan, John T. Jr., 'An almost absolute value in history' from Noonan, J.T. Jr. (ed), *The Morality of Abortion* (Harvard University Press, Cambridge, Mass., 1970), pp. 56–7.

11 Pluhar, Werner, 'Abortion and simple consciousness', *Journal of Philosophy* 74 (1977), 167.

12 The example comes from a letter by Brian Scarlett used, though in a different context, to argue against the views of Peter Singer and Helga Kuhse on the potential of the embryo: *Journal of Medical Ethics* 10 (1984), 217–18.

13 Strasser, Mark, 'Noonan on contraception and abortion', *Bioethics* I, 2 (April 1987), 199–205.

14 Grudzinskas, J. and Nysenbaum, A., 'Failure of human pregnancy after implantation', *Annals of the New York Academy of Science* 442 (1985), 39–44.

15 Ibid.

16 Roberts, C. and Lowe, C., 'Where have all the conceptions gone?', *Lancet* 1 (1975), 498–9.

17 Muller, J. et al., 'Fetal loss after implantation', *Lancet* 2 (1980), 554–6.

18 Braunstein, D., 'Chorionicgonadotrophin (HCG) and HCG-like substances in human tissue and bacteria' in J. Grudzinskas, et al. (eds), *Pregnancy Proteins: Biology, Chemistry and Clinical Application* (Academic Press, London, 1982), pp. 39–49.

19 Mahadevan, M. and Baker, G., 'Assessment and preparation of semen for *in vitro* fertilization' in C. Wood and A. Trounson (eds), *Clinical In Vitro Fertilization* (Springer-Verlag, New York, 1984), pp. 99–116.

20 Personal communication from Dr Ismail Kola, Centre for Early Human Development, Monash University.

21 This point is made by Senators Rosemary Crowley and Olive Zakharov in their dissent to the report of the Senate Select Committee on the Human Embryo Experimentation Bill 1985, *Human Embryo Experimentation in Australia* (Senator Michael Tate, chairman) (AGPS, Canberra, 1986); see especially paragraph D20.

9 | Arguing from potential

STEPHEN BUCKLE

One of the more common arguments employed in attempts to determine how we should treat the earliest forms of human life is the argument from potential. Typically, the argument holds that we should not interfere with, and perhaps should even assist, the development of the human fertilized egg because it already possesses the potential to be a fully self-conscious being; to be, that is, not merely a biologically human being (a human object), but a human subject.[1] It is, potentially, just like us, so we cannot deny it any rights or other forms of protection that we accord ourselves. The argument holds that, although the fertilized egg is not just like us in possessing rationality, self-consciousness, etc., this is a difference which is not fundamental, because it is overcome in time through the normal course of events. The normality in question is not statistical but biological, and is a matter of the proper functioning of biological processes. Proper functioning, in its turn, depends on the presence of an appropriately sympathetic environment.[2] The fertilized egg is not 'just like us' only in the sense that it is not *yet* just like us. Therefore, the argument concludes, we should not interfere with its natural development towards being a rational, self-conscious being. On its strongest interpretation, the argument is thought to establish that we should treat a potential human subject as if it were already an actual human subject.

The argument has not been without its critics, and some of the criticisms will be considered in this chapter. At least some of the criticisms, and the replies to them, give the impression that the argument's protagonists and critics are somewhat at cross-purposes. One aim of this chapter is to help to show why that should be so. The central aim, however, is to provide an analysis of some distinct ways of arguing about potential, and the different senses of 'potential' on which they rely. Given the attractiveness of this sort of argument to so many who think about the moral issues raised by the sophisticated new technologies available for intervening in human development from the earliest stages of (biologically) human life, this is an important endeavour. To give an example of this kind of argument, and to illustrate its attractiveness across a broad spectrum, I shall begin by giving a brief account of the viewpoint arrived at by the majority of the Australian Senate Select Committee commissioned to examine the question of experimentation on human embryos.[3] (The views expressed in the minority dissenting report of the same committee will be considered below.)

The majority report of the Senate Select Committee (the Tate Report) identifies, as the 'correct query' at the centre of its enquiries, the question: what is the respect due to the human embryo? (para. 2.40). It implicitly transforms this into the similar (but not identical) question: what features of the embryo (if any) command respect? To this question it offers the following answer:

> *It is in its orientation to the future that the Committee finds the feature of the embryo which commands such a degree of respect as to prohibit destructive non-therapeutic experimentation [para. 3.6].*

To the further question, What is this orientation, and why does it matter?, the report offers this answer:

> *If, as is the view of the Committee, the embryo may be properly described as genetically new human life organised as a distinct entity oriented towards further development, then the stance and behaviour proper to adopt towards it would include not frustrating a process which commands respect because its thrust is towards the future development of a biologically individuated member of the human species [para. 3.7].*

Although the report does not specify exactly what it is for a being to 'command respect', its reasoning and its conclusion are clear enough. Given the embryo's orientation towards the future, and given that there are no sharp discontinuities in embryonic development—and, therefore, nothing which could 'bear the weight' of being a 'marker event' for distinguishing the different moral values of different stages of development (para. 3.9)[4]—the report concludes thus:

> *In this situation prudence dictates that, until the contrary is demonstrated 'beyond reasonable doubt' (to use an expression well known in our community), the embryo of the human species should be regarded as if it were a human subject for the purposes of biomedical ethics [para. 3.18].* [Emphasis added.]

This, then, is the viewpoint of the majority report. Despite the fact that it employs some novel terminology, it clearly appeals to a version of the argument from potential. And, as do many versions of the argument, it concludes that we should treat some entity as if it were already something which it is not yet. It reaches this conclusion by considering some feature(s) which the entity already possesses, and which will help it to become that sort of entity which it is not yet. In the terminology of the report, this feature is, in the embryo's case, called its 'orientation to the future', an orientation which reflects its organization (paras 3.6–7). As we shall see below, this way of putting the matter is not without its virtues. However, having provided an account of a recently advanced version of the argument from potential, we can now consider a very general (and very dismissive) response to the argument.

An attempted dismissal

The argument from potential can be characterized as holding that the entity's potential for acquiring morally significant characteristics is itself morally significant. Putting the matter this way, however, immediately seems to show a difficulty: the argument implicitly concedes that the entity possesses no morally significant characteristics other than its capacity to come to have them. But how can the capacity to develop morally significant characteristics be itself such a characteristic? Of course, it cannot, at this level of generality, be shown that this capacity cannot be such a characteristic, but it does seem as if we are being asked to regard the capacity to be a particular kind of thing as *itself* one such thing. But this seems patently false. The argument from potential seems, therefore, to depend on a fundamental confusion— on regarding the potential to possess a certain feature as itself such a feature. To take a concrete example, it apparently regards a potential person as a particular kind of person. But whatever a potential person is, it is *not* a person. (This is not controversial. A child is a potential adult. Because it is a potential adult, it is not an actual adult. That is, it is not an adult.) So it seems that the argument from potential confuses the capacity for developing certain features with their actual possession. If so, it is a failure.

It might be thought that this negative conclusion can be avoided by denying that capacity and actuality are confused in the argument. But it seems that denying the existence of confusion leaves us no better off. For if the potential to develop morally significant characteristics is *not* such a characteristic, why should it be heeded? That is, wouldn't the argument then reject precisely what it seeks to establish, by conceding that the entity in question possesses, as yet, no morally significant characteristics at all? Why, then, should one treat it as if it did? It seems that, on either version, the argument must fail.

It is not unusual for variants of this argument to be advanced against the argument from potential, and there is no denying that it raises a number of awkward questions. However, there are two kinds of reasons which could be advanced in support of the view that, despite the dilemma just pointed out, the capacity to develop morally significant characteristics is indeed morally significant. They are, first, because it could be argued that it is precisely capacities, in particular the capacities of individuals, which *are* morally significant, that moral behaviour is a matter of respecting the capacities of other beings; and, second, that because present capacities are (in the ideal case) future actualities, and because it could be argued that moral behaviour is behaviour which regards the consequences of actions, then our moral obligations regard not merely present actualities but also future actualities, and therefore present potentialities.

Kinds of potential

The two kinds of reasons given above reflect two different kinds of moral theory. It is thus not surprising, although it does not appear to have been noticed, that they also indicate two different interpretations of potentiality and the difference it makes. Briefly, the first reason holds that respect is due to an existing being because it possesses the capacity or power to develop into a being which is worthy of respect in its own right; and respect is due to such a being because it is *the very same being* as the later being into which it develops. The already-existing being is a being which has the potential *to become* a being worthy of respect in its own right. The second kind of reason focuses on future outcomes, and so accords moral significance to whatever has the potential *to produce* certain states of affairs. By focusing on different sorts of potential, the two kinds of reason also generate two distinct versions of the argument from potential. These will be given separate treatment below.

To avoid misunderstanding it is important to indicate how the two kinds of potential differ. Both are central cases of potential, in as much as both refer to a power in (the *potency* of) an object or group of objects. This is well worth mentioning, since a number of recent accounts of potential regard it purely as a provisional prediction about future actualities. John Harris's is a case in point. He says, for example:

> To say that the fertilized egg is a potential human being is just to say that if certain things happen to it (like implantation) and certain other things do not (like spontaneous abortion) it will eventually become a human being. [5]

Although it is true that we sometimes employ the notion of potential in this very broad sense, it is not true that it is this sense which is employed in potentiality arguments about how to treat early human life. This can be readily shown. In this very broad sense, potentiality is just possibility—whatever the entity in question can possibly become, or be transformed into, is its potential. An acorn is in this sense not merely a potential oak tree; it is also potential food, or humus, or whatever human ingenuity can make of it. In like manner, a fetus is not only a potential person, it is also a potential abortus, a potential experimental subject, even a potential meal for the dog. Because any entity has, in this very broad sense of the term, so many different potentials, and these potentials are concerned with states of affairs of sharply contrasting value, we can conclude that this sense is too undiscriminating to found, without further qualification, any *moral* argument.

This would be a comparatively uninteresting observation if the argument from potential were an argument, not about the moral significance of the mere possession of potential, but about fostering a particular potential at the expense of other potentials. The argument is not, however, typically (if ever) of this form. As the above quotations from the Tate majority report help to indicate, the argument is typically couched in terms of allowing a process to run its natural course, not in terms of selecting between different potentials. This is no accident. If the argument were to be understood as concerned with selecting between different potentials, this would presumably amount to its being concerned with selecting between different outcomes. As such, it would be a straightforwardly consequentialist argument. This is surely not the *intention*, at least, of most of those who call on the argument; and the intuition that the argument is of quite a different colour finds some support in the fact that, if it were *simply* a consequentialist argument, then the notion of potentiality would be doing no crucial work. The argument could be rephrased without referring to potentiality at all. (There is, as has been indicated, a consequentialist version of the argument which depends on a narrower notion of potential. As a moral argument, however, its fate is not markedly different, as will be shown below.) So we can conclude that the argument from potential is typically presented as an argument about the moral significance of the mere possession of potential and, as such, if it is to be a viable argument, requires that potential cannot be understood simply as possibility.

It has already been pointed out that there *is* an available notion of potential which is appropriately narrower: the potential of an entity is its power to develop in certain ways, or to produce certain outcomes. It has also been claimed that this is the central meaning of the word 'potential'. This claim can be defended as follows. The use of 'potential' to mean possibility is most plausibly explained as the result either of broadening the conception of an entity's potency so that it includes the effects of that potency, or as a simple confusion of causes with their effects. Both of these alternatives reflect the fact that an entity's potency is related to its possibilities as cause is to effects: if an entity has a particular potency, then it has a particular possibility, or set of possibilities, which cannot be actualized except through the expression of that potency. It may be that, in some contexts, broadening the notion of potential may have its uses, and failing to discriminate between causes and their effects may not be harmful, but in the context of thinking about the status of early human life and the difference its potential might make shifts of this kind have greatly hindered enlightenment. I think it is not unfair to say that both advocates and critics of the potentiality argument are frequently not clear about what they understand potentiality to be, and that discussion of the argument has suffered from not distinguishing causes from their effects. (An example is provided below, in the discussion of the dissenting minority report.) As a result, discussion of the argument frequently seems to be at cross-purposes, or fails to scratch in the different places where potentiality has caused many moral consciences to itch.

Once it is recognized that the central meaning of 'potentiality' is that of an object's potency (the power it possesses in virtue of its specific constitution), and that arguments to moral value from potential invoke this central meaning, however unself-consciously, it can be seen that criticism of the argument along lines of the kind employed by John Harris will not succeed. From the standpoint of the central, relevant, meaning of 'potential', Harris's account does confuse the effect with its cause. Judgements of potential, in this relevant sense, indeed *imply* predictions (with some saving clause) about future actualities, but not because judgements of potential are provisional or conditional claims about the future. Rather, they are attributions of a present power to an entity, a power which will or can have that future effect.[6]

It is also worth noticing that Harris takes for granted that some form of identity is preserved in his brief account of potentiality: the fertilized egg is a potential human subject if *it* will eventually become a human subject, not merely because it will cause, or help to cause, a human subject to come to exist. More will be said on identity below, but Harris's assumption here is worth noticing in order to indicate that the concern of this chapter—with separating two notions of potentiality according to whether or not some form of identity is preserved—is not merely arbitrary. Assumptions about identity are intimately connected with the employment of the notion of potentiality. In particular, expressions of the form 'X is a potential Y' differ from all other forms of expression about potential precisely in the fact that expressions of this form require that a relevant kind of identity is preserved. (It hardly needs pointing out that expressions of this form are very common when the potentiality of early forms of human life is discussed.)

The two types of potential already referred to can now be more profitably considered. The potential *to become* is the power possessed by an entity to undergo changes which are changes *to itself*, that is, to undergo growth or, better still, *development*. The potential *to become* can thus be called developmental potential. The process of actualizing the potential *to become* preserves some form of individual identity.[7] It is for this reason that the potential *to become* is peculiarly appropriate to arguments which are concerned to establish the importance of respecting the capacities of a specific individual. The potential *to produce* differs in precisely this respect—it does not require that any form of identity be preserved; in fact, as we shall see, neither is its application limited to individuals. It is possible, therefore, to regard the potential *to become* as a special case of the potential *to produce*, and the latter as potential *simpliciter*. But no matter of substance depends on whether or not this particular course is followed.

In the following two sections, the two kinds of potentiality argument will be spelt out, and their implications determined. It should be stressed that the value of these arguments will not be considered; it will not be argued that the conclusions of either are true (or false, for that matter). My strategy is, more simply, to show what the arguments are, and what they imply; in particular, it will be argued that the arguments do not imply what the Tate majority report concludes. This obviously points up a problem with the report. But it should also be stressed that the following analysis, if correct, does not show the majority report's conclusions to be wrong; rather, it shows that its conclusions need different supports if they are to be credible.

The 'respect for capacities of individuals' argument

The first possible reason for holding that the capacity to develop morally significant characteristics is itself morally significant is that, according to one prominent kind of moral theory, respect for certain capacities of other beings is precisely the stuff of which morals is made. Of course, just *which* capacities (and, therefore, also just which other beings) do count for moral status is the crucial matter to be answered before this kind of approach can provide a concrete moral viewpoint. Most commonly, theories of this kind focus on the interests of sentient beings, thereby restricting the moral domain to such beings. Much broader versions have been offered, however, especially in recent times, with some environmentally concerned philosophers prepared to argue the case for trees or even ecosystems. To include the human conceptus, a broad version is certainly necessary. But it is not necessary, for the purposes of this chapter at least, to settle whether such a version (let alone one that defends the case of trees, etc.) is plausible. We will be concerned only with what such a view, if plausible, implies. Therefore no attempt will be made to settle the question of what capacities *do* count, morally speaking; in what follows, I will speak indifferently of the capacities of other beings, and leave the proper account of these capacities to one side.

It is also worth noticing that respect for the capacities of other beings may take two forms: it may be interpreted to require merely not interfering with the (non-harmful) exercise of capacities, or it may take a stronger form, such that mere non-interference is not adequate; that, at least in some circumstances, positive assistance is required. The difference between these interpretations, especially in the context of debates about IVF, is an important one. For, although the weaker version finds wide support as a plausible and practicable moral principle, the stronger is equally widely regarded as too onerous to be morally obligatory, and too conducive to the erosion of civil liberties. (To say this is not, of course, to settle the question of their plausibility as moral principles—but that is too large a question to be adequately treated here.) It is the stronger version which must be established if this general principle is to have implications for the very possibility of *in vitro* research: not interfering with an embryo in a petri dish is no aid to its future survival. The notion of guardianship, to which the majority report appeals (see paras 3.34 and 3.44 in particular) requires the stronger of the two interpretations. However, even the weaker version is sufficient to prevent destructive non-therapeutic experimentation, so the conclusions reached by the majority do not depend on the specific interpretation of the principle.

For those moral viewpoints which regard the rights of individuals as the fundamental factors of moral significance, respect for the capacities of other beings is the core of moral behaviour. Rights are commonly explained in terms of the respect due to beings which possess interests, and interests in their turn reflect a set of capacities—whether to suffer harms, or to engage actively in pursuing certain goals or other course of action. Since a potential being is a being with certain capacities, it is perhaps not surprising that the argument from potential is so frequently couched in terms of rights. In so doing the argument attempts to establish that the early embryo, in virtue of its developmental capacities, is a worthy object of respect because it is a being with the potential *to become* a human subject.

It is necessary to employ these rather awkward formulations in order to make clear that this form of the argument from potential depends on identifying a particular being which is worthy of respect in virtue of what it has the potential *to become*. We must, on this version, be able to identify that individual which will itself come to be the human subject. This version of the argument, therefore, does not apply where such an identification is not possible—either because there is simply no such individual, or even because there is *not yet* such an individual. It is important to recognize this fact. It has a significant bearing on this, the individual respecting, version of the argument.

The argument from potential is commonly deployed to advance the claims of the fertilized egg as an entity worthy of respect because of its potential. The argument, thus deployed, is what has been described here as a 'respect for capacities' argument, and as such it discriminates sharply between the status of the fertilized egg and its progenitors, the unfertilized egg and sperm. It has been criticized for doing so. The objection runs as follows: if we can identify a particular sperm and the egg it will fertilize, why cannot we say of them that together they have much the same potential as the fertilized egg? After all, provided fertilization is successfully achieved (and new IVF techniques may provide a high success rate in the laboratory), are they not (almost) as likely to generate a human subject as is the fertilized egg? As far as likely outcomes are concerned, the objection is quite pertinent. However, the defender of the potentiality argument is likely to feel that such considerations miss the point; and the above exposition of the argument shows that this is indeed so. The potentiality argument, thus employed, is not about likely outcomes, nor even about outcomes at all— it is, rather, about respecting a being which is worthy of respect. Neither the sperm nor the egg are beings worthy of respect, on this view, because neither has the potential *to become* a human subject, even though both of course have the potential to help to produce a human subject. So this version of the potentiality argument appears to be justified in discriminating between the fertilized egg and its progenitors. (The possibility of parthenogenetic development of the unfertilized egg is a problem for this view, however. How serious a problem it is will not be considered here, since it involves complications which lead too wide of my main purpose.[8])

It might be objected at this point that, although the sperm and egg do not individually have the potential *to become* a human subject, when considered together, they do. Arguing thus requires that we allow that a potential entity need not be composed only of a single object, and there seem to be reasons for thinking this to be quite legitimate—we can justly speak of the potential of an army, or of a sporting team, for example. Therefore, it might be argued, although separate objects, the sperm and egg can be legitimately regarded as jointly constituting an entity with the potential *to become* a human subject.

The argument will not work, however, because it is misplaced. As an argument about an entity which has the potential *to produce*, it seems quite acceptable, and will be considered in the next section. However, it will not do as an argument about the potential *to become*, since that kind of potential attaches only to distinct individuals which preserve their identity over time. Therefore it attaches only to entities which, if they are composed of distinct parts, nevertheless can be classed as a distinct single individual. To satisfy this condition, the several and distinct parts must in some way constitute a complex *whole*. This is true of armies and, to a lesser extent, of sporting teams—although composed of discrete individuals, armies or sporting teams are (fighting or sporting) *units*—they have a unity constituted by a specific organization directed towards a particular end. (It should be noted that what constitutes some group of objects as a complex whole will, in all probability, vary in different cases. This need not trouble us, however; the central question will always be whether such a whole, however constituted, preserves its identity through whatever changes it undergoes.) Consequently, armies and sporting teams can be understood to possess the potential *to become* in all those contexts where they are considered as complex wholes, that is, wherever they are not being considered merely as a collection of discrete human beings.

Where a collection of discrete entities is not organized into a whole, there is no individual to possess the potential *to become*, no individual which develops through the actualization of the potential. So, where there is no unified whole, there is no potential *to become*. No such restriction exists if we are considering the potential *to produce*. A mixture of gases, if it includes hydrogen and oxygen, has the potential *to produce* water, despite the fact that such a mixture has no organizational unity. So, although it is possible to speak of the potential of entities which are composed of a number of discrete objects lacking in any unifying features, the potential in question is the potential *to produce*. In contrast, it is only possible to speak of entities having a potential *to become* where such entities are complex wholes.

In the case of the sperm and egg, there is no complex unity, no overarching organization. Such unity or organization arises only with fertilization (in fact, only with the completion of the fertilization process—perhaps at syngamy). Prior to the event of fertilization, there are two organized, complex wholes—the sperm and the (unfertilized) egg. So the sperm and the egg, even when considered jointly, as a collectivity, fail to satisfy a necessary condition for the possession of the potential *to become*. So, although together they have the potential *to produce* a human subject, they do not have the potential *to become* a human subject.

The argument from potential, in its non-consequentialist 'respect for capacities' version, is that respect is due to a being which has the potential *to become* a human subject, because of that being's potential *to become* such a subject. The argument, therefore, applies only when there exists a being with this particular species of potential, and only *to* such a being. It does not apply to any being, nor to any other kind of entity (unified or not), which has the more general potential *to produce* a human subject. Therefore it does not apply to the sperm and egg, prior to fertilization. It is, however, commonly understood to apply to the fertilized egg, and, of course, to subsequent stages of the development of the human infant.

This common understanding is not beyond criticism. For, although it is certainly true that the fertilized egg stands at the beginning of a process of development which ends (in the normal case) in a human subject, this shows only that the fertilized egg has the potential *to produce* a human subject. To show that it also has the potential *to become* that subject it is necessary to show that the fertilized egg is the same being (albeit writ very small) as the human subject at the other end of the developmental process. There are good reasons for doubting whether this can be done, principally because the fertilized egg cannot be identified as the same being as the embryo proper, that is, it cannot be identified as the entity which begins to develop at about the fourteenth day after fertilization (later developing arms, legs, etc.), and which comes to be sustained by the life-support system provided by the uterus and placenta. The identification cannot be made because the changes that the fertilized egg undergoes are not changes through which it develops into, or itself becomes, the embryo proper. Rather, it undergoes a process of differentiation in which the various cells developed in the earliest stages after fertilization take on a range of different functions, only one of which is the development of the embryo proper; and no one group of specialized cells can be singled out as the same individual as the fertilized egg.

This is perhaps not widely understood, so it is worth spelling out in some detail. The reproductive biologist Anne McLaren describes the process as follows:

> *The first two weeks after fertilization are essentially a period of preparation for the later development of the embryo. The fertilized egg divides once or twice a day for the first few days, to form a clump of cells which then spends about the next week burrowing into the wall of the woman's uterus. During this period of implantation, most of the cells become progressively committed to various tasks concerned with the protection and nourishment of the future embryo. Eventually they or their descendants form the placenta and the various other tissues that surround the embryo—chorionic villi, secondary yolk sac, primary mesoblast, amnion, allantois, and so forth.*
>
> *At the end of the period of implantation, there remain some cells not involved in any of these life-support systems. It is in this group of cells (the 'embryonic plate') that the so-called primitive streak appears, marking the place where the embryo itself finally begins to develop.*[9]

The entity McLaren refers to here as the 'embryo' is what we have called the 'embryo proper'. It is not what the Tate Report refers to by its use of 'embryo', however—that is, the human conceptus from fertilizaton onwards. The two entities are not the same, and McLaren rightly stresses the confusion that results from failing to recognize this:

> *If this entity forming around the primitive streak is the embryo . . . what are we to call the entire collection of cells derived from the fertilized egg, of which the embryo is a tiny subset? In recent years, we researchers have developed the bad habit of calling the whole set of cells, at each prior stage, the embryo as well. To the non-specialist this is confusing, just as it would be confusing persistently to refer to the [British] Shadow Cabinet as the Labour Party or vice versa.*[10]

McLaren is quite correct, although she has, if anything, understated the point: it would be confusing to identify the Shadow Cabinet with the Opposition (parliamentary) party because, in any case other than a complete electoral disaster for the Opposition, it would simply be wrong to do so. The Shadow Cabinet is not the Opposition party, even though it is part of, or comes from, that party. Even though the Opposition party can be said to produce the Shadow Cabinet, it is not itself the Shadow Cabinet. The two are distinct entities.

The same is true of the human conceptus after fertilization and what we have called the embryo proper. The embryo proper is part of the organic system that develops from the fertilized egg, it comes from the fertilized egg, but it is not itself the same entity as the fertilized egg. The fertilized egg produces the embryo proper, but it is not itself the embryo proper. That is, the fertilized egg has the potential *to produce*, but not the potential *to become*, the embryo proper. But it is the latter, not the former, which is the kind of potential peculiar to individuals, and which is, therefore, the only kind of potential that can be plausibly regarded as capable of grounding whatever respect is due to an individual in virtue of its own capacities. Therefore the human conceptus, prior to the formation of the embryo proper, does not possess the kind of potential necessary for rendering respect for its capacities an appropriate response.

But if the human conceptus prior to the formation of the embryo proper does not itself possess the potential *to become* a human subject, and, therefore, does not possess the kind of potential necessary for it to be accorded recognition as worthy of respect for its capacities, then the individual-respecting version of the argument from potential does not apply to the fertilized egg, or to subsequent stages prior to the formation of the embryo proper. As it stands, then, this version of the argument from potential does not establish the conclusion it is commonly thought to establish.

To conclude this section, it should also be observed that the moral significance of this version of the argument from potential has not been established, nor has it been denied. All that has been determined is where the argument applies, and where it does not. Therefore, to argue, as above, that this version of the argument does not apply before the formation of the embryo proper is not to argue that where the argument does apply, it is morally decisive, or even significant. The discussion of this argument has concluded only that it cannot apply where there is no embryo proper; it has not concluded that where it does apply, it is decisive. The special moral significance of this form of potential, and, therefore, of the embryo proper, remains to be settled.

The consequentialist argument

The consequentialist version of the argument that the capacity to develop morally significant characteristics is itself morally significant is, fortunately, much less complicated than the 'respect for the capacities of individuals' version. As a result, it is also much easier to determine its implications. As already noted, this version of the argument focuses on the potential of an entity *to produce* morally significant states of affairs (commonly, but not necessarily, this state of affairs will be the existence of a future being), rather than on its potential *to become* a being of a certain kind.

On this version, then, tangled questions concerning whether or not there exists a being which will itself become the future being (mercifully) do not arise. For this reason, there is no important distinction to be drawn between the fertilized egg and the embryo proper. Both have the potential *to produce* a future human subject. (Whether they have the same potential, or whether it is possible to accord degrees of potential *to produce*, will not be considered here.) So it may seem that this version of the argument is more promising as a support for the special moral significance of the fertilized egg.

This, however, is not the case. The reason is that, from a consequentialist standpoint, there is no crucial moral difference between the fertilized egg and the sperm and unfertilized egg—considered as a pair—from which it is formed. The sperm and unfertilized egg, when considered jointly, also have the potential *to produce* a future human subject, even though that potential is not *activated* until fertilization occurs. So this version of the argument does not establish a crucial difference in moral status between the fertilized egg and its biological forebears.

Does this mean that the consequentialist version of the argument must conclude that we ought not to interfere with, perhaps even encourage the conjunction of, any specific sperm and egg pair? No, it does not, because this version of the argument attaches moral significance to the potential *to produce* (that is, to the presently existing capacities) only derivatively. The central locus of moral value is here, as it is for all consequentialist arguments, not present potentialities but future realities. It is to what will be produced: that is, to the future human subject. The present potential (*to produce*) bears moral weight according to the moral weight of the possible future human subject. This means that, although it has genuine moral significance, this significance is not of the kind sought in the 'respect for capacities' version of the argument, but that significance necessary to figure in calculations concerning best outcomes.

So, in the end, this version of the argument from potential is best understood as one form of an argument about the moral significance of possible or future human subjects. As such, it is thus not unlike the argument from mere possibility considered above. Although it focuses on specific individuals, the moral weight of the argument rests on the value of possible future states of affairs; values which are ascertained by comparison with other possible futures. This point can be more fully explained as follows. If allowing a present embryo to develop will produce a future state that is less valuable than preventing the development of this embryo, and developing instead another, not yet conceived, embryo (for example, if the present one is suffering from a congenital defect, or if the parents or other appropriately placed parties are unable to care adequately for it), then a straightforward application of consequentialist principles requires that we follow the latter course. This course, however, attaches less moral significance to an actual being than to a merely possible one because, in this case at least, the greater moral value is achieved thereby. What course should be followed, then, is not, on this view, determined by whether there is or is not an actually existing entity, but by determining the best outcome, regardless of whether this outcome is to be achieved by actual or merely possible entities.[11]

What all this means, as far as our practical purposes here are concerned, is that the consequentialist version of the potentiality (cum-possibility) argument has implications even more contrary to the conclusions of the majority report of the Tate Committee than the 'respect for capacities' version of the argument. For the consequentialist argument does not identify a point (a marker event) beyond which experimentation cannot be justified. This does not mean, it should be stressed, that it can be used to support experimentation willy-nilly. Far from it: what it means is that the justifiability of any particular research project is a matter of the perceivable costs and benefits likely to be achieved thereby, costs and benefits which will clearly include those to the developing fetus if it has reached a stage where it can be affected (and perhaps in other cases as well). Such projects will, therefore, justifiably involve more extensive experimentation on human embryos, provided the goals of such research are both sufficiently valuable and clearly achievable. Thus Dr Alan Trounson, in his submission to the Tate Committee, argued that 'if suddenly we get the answer to the whole of cancer or the whole of every debilitating disease by studying 200 28-day embryos', such study (presumably destructive) would be defensible (para. 3.16). Dr Trounson's argument here is clearly inspired by consequentialist considerations, and his argument serves well to illustrate the sorts of arguments possible within a consequentialist framework.

The point here is not whether or not such arguments are compelling. It is, rather, to give an idea of where consequentialist arguments may lead; and to underline the fact of their divergence from the conclusions arrived at by the Tate Committee's majority report. So both versions of the argument from potential imply conclusions different from those arrived at by the majority report. If the report's conclusions are not to be abandoned, it is necessary to provide them with some foundation other than that of potentiality.

It is perhaps necessary to answer one possible complaint about the method employed in this discussion of the potentiality arguments. It is surely cheating, it might be objected, to divide what is normally presented as one argument into two, and then to show that the argument fails because neither half of it gives the conclusion sought. Why cannot the argument be understood as some combination of these two sorts of common moral consideration (i.e. of consequences and respect for individuals), and as such a combination come up with the anticipated conclusion? The short answer to this objection is that perhaps the best version of this type of argument would indeed combine both sorts of consideration in some way, but that it is very difficult to see how it could establish the claims of the fertilized egg to possess the relevant potential in a way that excluded both its antecedents and consequents. It is difficult to see this because the combined argument would either recognize moral significance in the existence of the relevant kind of individual or it would not. But in the former case, there is no good reason for thinking that the relevant individual could be the fertilized egg rather than the embryo proper; and in the latter, there is no reason for focusing on existing individuals at all, and hence no reason for thinking that the fertilized egg will come to carry any moral weight denied it by the consequentialist version of the argument already considered. So there seems to be no reason for departing from the conclusions already established.

The dissenting report

To this point, I have been concerned to show two possible ways of construing the argument from potential, and to argue that neither version establishes the moral distinctiveness of the fertilized egg. In this section, I shall examine the report of the dissenting minority of the Tate Committee, because they also argue that, properly interpreted, considerations of potential do not indicate any special status for the fertilized egg. It may be thought, then, that this section will provide a defence of the dissenting opinion. This is not so; although I will argue that a revised version of it may bear fruit. As it stands, however, the minority report provides a good example of one of the ways of arguing about potential that this chapter has shown to be inadequate. Illustrating some of these inadequacies will be helpful, because doing so will serve to clarify the account of potential offered in this chapter, and to avoid some further pitfalls. It can then be shown how a more viable reconstructed argument is possible.

The minority report advocates, as the vital marker event, the stage of implantation in the uterus, a process which normally comes to completion around the fourteenth day after fertilization. Although this is a different criterion from that implied by the individual-respecting version of the potentiality argument, it refers to a similar stage of development. (The embryo proper begins to form as the process of implantation reaches completion.) Further, the minority report's defence of this proposal depends on no new principles. Like the majority report, it too appeals to potential to ground its conclusions. In somewhat similar vein to the individual-respecting version of the argument already considered, the minority report argues that prior to the proposed marker event there is in fact no potential at all, and, therefore, that the argument from potential does not apply prior to this event. So, like the analysis of the argument from potential provided above, the minority report seeks not to discredit the argument from potential, but to show that, properly understood, the argument gives rise to conclusions quite different from those reached in the majority report.

In this attempt, however, the minority report fails, because it depends on an inadequate notion of potentiality. In terms of the account of potential offered in this chapter, it confuses potential with the probability of that potential's actualization. It identifies the potential of an entity with that entity's probable future. It holds, for example, that 'developmental potential is dependent on successful implantation, or at least the opportunity for this to occur' (D.26). In the following discussion, this position will first be spelt out a little more fully. Then its weaknesses will be pointed out, and a contrast drawn with the account of potential provided in this chapter.

The identification in the minority report of the potential of a being and its probable future is best illustrated by the following passage:

> *Any object or thing has an infinite number of possible future courses. For a non-sentient or inanimate thing, e.g. a rock, the particular future outcome that actually happens is determined by forces outside itself. An embryo is like a rock in this respect—it cannot make decisions for itself. Its future is decided by others. It has potential only in virtue of decisions by others about it. If there is a clearly defined responsible party or parties their decisions determine the embryo's potential and that becomes the embryo's potential [D.20].*

The inadequacies of this approach, even for the purposes of the minority report, can be illustrated as follows. The argument provides no support for the view that implantation in the uterus is an event of moral significance, for even where such implantation is the consequence of a specific decision to that effect (as is so in the IVF case), new decisions can be made concerning, and can therefore modify the potential of, the embryo at any stage before it becomes capable of deciding for itself. Given that even newborn infants lack the capacity to choose for themselves, this account effectively denies potential a role in governing the decisions we make about fetuses (whatever their stage of development), and even of newborn infants. In all such cases, the potential of the being could be determined only by the decisions we make about it. The minority report thus removes any guiding role for arguments from potential. If, as that report seeks to establish, implantation is to matter as an event of moral significance, it must be for reasons other than the potential of the embryo.

This is sufficient to show that the relevant notion of potential has been lost. For the notion of potential employed in moral arguments is not a mere free rider. The potential of a being is, in all such arguments, advanced as a reason for acting in certain ways rather than in others: it is not a consequence of such actions. If we insist instead that potential is a power possessed by a being, we can neatly capture this feature of such arguments. An everyday example will serve to show this. If a child is declared to be a potential Mozart, this is a judgement about the child's talent (one of the more common terms in which potentiality judgements can be expressed). If the child's parents decide to discontinue the piano lessons necessary for becoming a Mozart, the child's potential is no more affected than is its talent. Its potential is, however, a reason for the parents not to discontinue the lessons. The lessons (and, indirectly, the parents' decisions about them) are important because of the potential they serve to realize; not because they constitute, or even partially constitute, that potential. The minority report's account of potential, and the anomalous results to which it gives rise, can thus be construed as a confusion of causes with their effects.

If we now return to the question of implantation, we can see that, on the account of potential provided in this chapter, it is not potential but the likelihood that potential will be actualized that is dependent on the decision to implant. It is *because of* the potential of the entity that implantation is a crucial factor in its development. Acorns do not become potential oak trees when they are planted; rather, because they *are* potential oak trees, when they are planted they begin the journey towards being actual oak trees. As previously stressed, the potentiality of an entity is not what it will probably become, but the power it has to become something, whether or not that becoming is probable. An entity is a potential X if it has the power to become X; that is, if it will become X in virtue of the operation or expression of properties of its own, given circumstances conducive to the operation or expression of those properties. If the relevant circumstances do not apply, then of course it will not become X, but this shows only that its potential will be frustrated (will not be actualized), not that it lacks such potential. The possibility of frustrating an entity's potential reflects the fact that being a potential X is not a sufficient condition for becoming an X.

The minority report's account of potentiality is clearly inadequate. However, this does not mean that its substantive position likewise fails. If anything, it can be defended rather more persuasively once the erroneous account of potentiality is set aside. This is so because the dissenting report's conclusions, or other conclusions not markedly dissimilar, can be rebuilt on more adequate grounds. Rebuilding thus is possible because this report implicitly rejects the individual-regarding version of the potentiality argument by focusing on decisions about the embryo rather than on questions of the embryo's own status. Its basic viewpoint can, therefore, be reasonably captured by some version of the consequentialist argument: it is perhaps best understood as an attempt to determine the scope, and proper bearers, of responsibilities for possible future human subjects. Thus understood, the dissenting report can be detached not only from its explicit dependence on a defective notion of potentiality but also from any appeal to potentiality at all, as the account of the consequentialist version of the argument has shown. The dissenting viewpoint could then be reconstructed along the best available consequentialist lines, and its plausibility then reassessed.

Concluding remarks

This chapter has attempted to distinguish between different ways of thinking about potential, and different ways of arguing from the senses thus distinguished. It has applied the results gained to the conclusions drawn from potentiality arguments by the Australian Senate Select Committee's report, *Human Embryo Experimentation in Australia*. It has been argued, firstly, that neither version of the argument supports the common belief, which is also the view of the majority report, that considerations of potentiality establish the special moral status of the fertilized egg. Secondly, it has been shown that an alternative employment of the argument in the dissenting report utilizes a notion of potentiality which is seriously inadequate, but which can be reconstructed as a consequentialist argument from possibility.

If the general thrust of the discussion of potentiality in this paper is defensible, then the philosophical debate about arguments from potential is deficient in two related respects: the relevant notion of potentiality is frequently misunderstood, and, as a result, some fundamental differences of interpretation—reflecting deep-seated commitments to either deontological or consequentialist moral thinking—go unrecognized, thus serving to obscure the precise nature and source of the disagreement. It is perhaps too much to hope that this will actually help to settle those disagreements; but it should serve to illuminate some previously unrecognized strengths and weaknesses of the more common positions.

Notes

1 This chapter employs the term 'human subject' to refer to a human organism of moral and therefore legal considerability. The term has been chosen in an attempt to be agnostic about contentious issues.

2 For some suggestive remarks to this effect, I am indebted to an unpublished paper by Michaelis Michael, 'The moral significance of potential for personhood'.

3 Senate Select Committee on the Human Embryo Experimentation Bill 1985, *Human Embryo Experimentation in Australia* (Senator Michael Tate, chairman), (AGPS, Canberra, 1986). (The report employs the term 'embryo' to refer to the earliest stages of human life, including the just-fertilized egg, in contrast to much recent scientific practice.)

4 The conclusions of the majority report thus depend not merely on the argument from potential, but also on the view that fertilization is the only decisive event in the course of the development of the human infant. But cf. my 'Biological processes and moral events' (Chapter 17 of this book); Bernard Williams, 'Which slopes are slippery?' in Michael Lockwood (ed.), *Moral Dilemmas in Modern Medicine* (Oxford University Press, Oxford, 1985), pp. 126–37; and 'Types of moral argument against embryo research' in CIBA Foundation, *Human Embryo Research: Yes or No?* (Tavistock, London, 1986), pp. 185–94.

5 Harris, John, '*In vitro* fertilization: the ethical issues', *Philosophical Quarterly* 33 (1983), 223.

6 The power in question is, as the preceding discussion has implied, a causal power. For an elaboration and defence of this notion see R. Harré and E.H. Madden, *Causal Powers* (Basil Blackwell, Oxford, 1975).

7 The identity which is preserved in such cases is not personal identity but *numerical* identity—the potential and actual beings are identical in that they are the same

single thing, albeit at different stages of development. See Derek Parfit, *Reasons and Persons* (Clarendon Press, Oxford, 1984), ch. 10.

8 On these issues, see also Peter Singer and Karen Dawson, Chapter 8, 'IVF technology and the argument from potential'.

9 McLaren, Anne, 'Why study early human development?', *New Scientist*, 24 April 1986, 49.

10 Ibid.

11 This approach is exemplified in the work of R.M. Hare. See, most notably, 'Survival of the weakest' in Samuel Gorovitz et al. (eds), *Moral Problems in Medicine* (Prentice-Hall, New Jersey, 1976), pp. 364–9.

10 | Who is the subject of the research?

BETH GAZE AND KAREN DAWSON

The debate over human embryo experimentation in recent years has centred on whether or not destructive embryo research is acceptable. In this debate, the meaning of the term 'destructive experimentation' is rarely specified: it is assumed to be clear. The debate has taken place in the context of a broader system of regulation of research on human subjects which distinguishes 'therapeutic' and 'non-therapeutic' experimentation from medical practice. Again, it is assumed that these distinctions are clear and well understood. In this paper we examine one of the difficulties of applying these concepts to *in vitro* human embryo research by asking the question: who is the subject of the research? We do not discuss the moral status of the embryo or the morality of embryo research. The question of who is the subject raises problems regardless of the moral status attributed to the embryo. We begin by examining the background concepts of therapeutic, nontherapeutic and destructive experimentation in medical practice and research on human subjects generally and the problems of applying these concepts to human embryo research.

The fields of medical practice and biomedical research, previously largely left to the control and concern of those working in them, have in recent years been regulated in two ways: ethical regulation and legislation. Ethical regulation of human experimentation (often referred to as 'self-regulation') developed from the principles enunciated after World War II by the Nuremberg court which tried German doctors for crimes against humanity. International guidelines for research on human subjects are given by the Declaration of Helsinki, formulated by the World Medical Association in 1964 and revised in 1975.

In this chapter we focus on one aspect of the adequacy of an ethical regulatory system for regulating embryo research. Because we avoid discussing the moral status of the embryo, we shall not consider the question of whether the human embryo is a subject to which the ethical regulatory codes (including the Declaration of Helsinki) apply. We approach the issues instead by asking: who is the subject of embryo research? We consider some of the problems which arise from the failure of ethical guidelines to take account of the factual context of the dependence of embryos on women.

We use examples based on the Australian experience to illustrate the argument. Problems arising in Australia are typical of problems which may also arise in other countries. The wider significance of the Australian experience lies in its providing one of the few existing models of the interaction of legal regulation of *in vitro* fertilization practice and embryo research with ethical regulation. For the purpose of this chapter, however, our focus is on the ethical, rather than legal, regulation of the area.

Both ethical and legislative regulation have been adopted in Australia. The National Health and Medical Research Council (NH & MRC) set out a code regulating research on human subjects in its Statement on Human Experimentation, first adopted in 1965 and last revised in 1985. Supplementary Notes to the statement have been introduced on several occasions since 1982 to expand its guidance in specific areas. Legislative regulation of medical practice and research has not been comprehensive, but has been introduced as necessary to deal with particular areas of controversy. The practice of IVF and the conduct of IVF research using human embryos have been regulated by legislation in some states.[1]

We begin by considering the general application of the regulatory codes, which depends on classifying activities into different categories. This can be difficult, because the meanings of the terms used to describe or define the categories are ill-defined. Problems of ambiguity and lack of clarity in basic concepts pervade the area. For example, whether the NH & MRC code (or any other regulatory code) should be described as self-regulation depends on whose perspective is adopted. From the viewpoint of IVF practitioners and researchers, the code is externally imposed by the NH & MRC through its Medical Research Ethics Committee, whose membership is not drawn exclusively or even mainly from those involved with IVF. From the perspective of the general community, the code is imposed by the principal medical research funding and monitoring body in Australia on medical researchers and, therefore, is self-regulatory. Nor is it clear that the NH & MRC code should be described as ethical regulation: it creates a mechanism for the application of ethical principles rather than specifying the ethical principles which should be followed. The Statement on Human Experimentation lays down in general terms the basic principles of research on humans, such as the need for informed consent and scientific value. Although the Supplementary Notes lay down further principles in specific areas, they are still quite general. The major mechanism by which the NH &

MRC code is put into effect is through the Institutional Ethics Committees required by Supplementary Note 2, which creates a mechanism for the application of the general principles to specific research proposals. The procedural or institutional requirements of the code should be distinguished from moral or ethical argument about which substantive principles should be applied. In this chapter we refer to systems of 'ethical self-regulation' such as those of the NH & MRC and World Medical Association as regulatory codes, to avoid the problems of labelling them as self-regulatory, or ethical regulation.

Distinctions in the biomedical field

The introduction of a regulatory code to govern biomedical research means we must identify where practice ends and research begins. If different types of research are to be treated differently under the regulatory codes, we must be able to clearly identify and classify them. The need to make these distinctions, therefore, flows from the adoption of regulatory codes for medical research. (The source of these distinctions, the Declaration of Helsinki, is discussed further below.)

The standard concepts of practice and research are quite simple. In general, medical practice involves the use of existing knowledge to treat a patient, while research involves the testing of hypotheses, which may reveal new knowledge. The significance of the distinction is twofold. First, more stringent requirements for informed consent by the subject apply to research than to medical practice. Second, in research on human subjects, care must be taken to ensure that the potential benefit of the research is not outweighed by the present risk to the subject.

Research on human subjects is often differentiated according to whether or not it is for the benefit of the subject, that is, whether or not it is therapeutic research. On this approach, research which will not benefit the particular subject but which will produce results for the direct benefit of other sufferers of the same condition, or even for the later benefit of the subject, is regarded as non-therapeutic. Ethically, research which is not intended to benefit the subject requires more justification, especially if it involves a risk of harm to the subject. Therapeutic research raises special problems concerning the consent of the subject because it takes place in the context of the doctor–patient relationship.

While we may be able to identify some procedures which are clearly accepted as medical practice, or clearly therapeutic or non-therapeutic research, many procedures lie in the grey area between, and to classify them we need better criteria. The problem can be clarified by distinguishing its theoretical and practical aspects. First, the theoretical concepts involved and the distinctions between them must be stated. Second, to have an idea of the practical effect of the distinctions, we must look at actual day-to-day practices and see how they correspond with or depart from the theoretical statement. There is, however, a third complication to be considered: the simple statement of the conceptual distinctions fails to take account of changes over time, and this dimension is crucial. Given that neither medical practice nor biomedical research are static, the process by which procedures move over time from being experimental to being innovative treatments and to becoming accepted as part of orthodox medical practice is of vital importance in relation to regulatory codes. The time dimension is also crucial to the process of planning research, because decisions about the ethical acceptability of research have to be made without the benefit of hindsight. The judgement whether or not a research project is ethically acceptable and should be permitted must be made on the basis of a proposal before the project is undertaken, not after it has been completed in the light of its success or its failure to provide the anticipated results.[2]

Distinguishing medical practice and biomedical research

Taking the field of biomedical activity as a whole, should we begin by identifying 'research' or by identifying 'practice'? Although the regulatory codes and existing practice in the field suggest that it is research which should be identified, it is probably easier (and theoretically preferable) positively to identify medical practice, defining research negatively as whatever is not practice. The question then becomes how to identify what is within the field of accepted medical practice. Various possibilities exist: we could accept the opinion of the person proposing the procedure, or the evidence of other medical practitioners that they regard it as established medical practice, or we could seek 'objective' evidence of the validation of the procedure by evaluating it against the background of existing knowledge in the biomedical field including its past record.[3]

Three areas of uncertainty must then be dealt with in distinguishing practice from research: the identification and scope of the 'class of activities'[4] in question; the method of validating procedures as part of medical practice; and the routine uncertainty of medical practice.

Formulating the activities

How widely or narrowly can a 'class of activities' extend? Must a proposed procedure be considered as a whole, or is it possible to consider it in parts, some of which may be already established? Can different procedures be aggregated in one research proposal? In the absence of guidance as to the scope of a 'class of activities', the classifications can be manipulated. Such problems of characterization can occur in connection with IVF research.

Validation

The customary standard for routine practice is a 'reasonable expectation of success'. In this context validation must be found in the results of previous research or clinical application which provide a basis for the reasonable expectation. Such an expectation can only be formed by people with adequate knowledge, so validation must be seen as a matter of professional opinion. This may raise problems of proof if there is a significant difference of opinion within the profession. It indicates that the opinion or therapeutic intention of the practitioner proposing the treatment is not solely determinative, although in reality it must be the starting place.

The recent inquiry in New Zealand into the failure by doctors at the National Women's Hospital to treat women returning a positive test for cervical cancer on Pap smears provides a clear example of the problems inherent in this approach.[5] Because a powerful senior doctor at the hospital took the view that the cancer *in situ* disclosed by the Pap smear was not a precursor to invasive cervical cancer, the women involved were not given the orthodox treatment for this condition. The Committee of Inquiry concluded that the project was a research project, although the proposal, which had been considered at a Hospital Medical Committee meeting in 1966 (before an ethics committee was set up at the hospital) was inadequate, and requirements for research such as obtaining consent from patients were not observed.[6] That the doctor could not only begin but also continue such a project for decades as if the treatment were validated, in the face of dissent from colleagues and strong contrary evidence from overseas research and clinical experience, suggests clear problems in the process of scrutinizing claims to validation as well as in approving research projects. Within a hierarchical profession such as the medical profession, differences over validation may be difficult to separate from the status of the disputing groups.

One alternative to the informal process of publication of results followed by practitioners' adoption of the new procedures is a formal accreditation process. Some examples exist, but only in limited areas. For example, in some hospitals, adoption of new clinical procedures may be formalized through ethics or practice committees. Government schemes regulating the use of, or public funding for, drugs provide a second avenue for formal acceptance, but only for drug therapies.[7] In Australia, IVF has been treated as a special case by the NH & MRC in its Supplementary Note on *in vitro* fertilization and embryo transfer (IVF and ET), which requires institutional ethics committee (IEC) approval of all aspects of a clinical program.[8] This confers on the IEC the function of accrediting the IVF and ET procedure for clinical application in addition to its normal function of vetting research proposals. The note contains no guidance to IECs as to the criteria to be applied in assessing suitability for clinical application, but presumably the usual criterion of validation would continue to apply. For clinical IVF, however, validation is not available, and this will always be the case when a research or experimental procedure makes the jump into practice.

The price of formalizing accreditation of new procedures or activities for practice is reducing the flexibility of medical and therapeutic innovation. Some reduction in flexibility is probably an unavoidable consequence of the imposition of any formal regulation of biomedical research. Where regulation depends on classifying procedures, the method used for determining whether a procedure is within or outside the regulated field becomes very important. Given that such regulation exists to protect the subject of research, a balance must be struck between that protection, and the reduction in flexibility of innovation in medical practice and research (which also benefits patients generally). Adopting more formal accreditation procedures accords more weight to the protection of the subject, as well as requiring clarification of the research/practice divide.

Treatment variations

Medical practitioners, like other skilled workers who deal with individuals, are expected and required to vary their therapeutic practice to take account of the individual characteristics and reactions of their patients. Some variation is, therefore, inevitable in medical practice, and no external regulatory approval is required for individual variations when applying existing or orthodox treatments.[9] The difficulty is that the limits of the medical practitioner's therapeutic discretion are uncertain, and it may at the edge shade into the conduct of therapeutic research. Although the dichotomy between therapeutic and non-therapeutic research has been criticized, it still persists.

When does treatment variation become significant enough to attract attention as 'research' with the need for approval by IECs? The major problem is the question of degree. Some examples may be clear: application of new drugs, or of new surgical procedures. But what about cases where a drug is used at different levels or for different purposes,[10] or where variations are made in the culture medium used to maintain embryos prior to their transfer to a woman undergoing IVF? What degree of variation converts the case from one of clinical practice to one of therapeutic research? Guidelines in regulatory codes issued by bodies such as the NH & MRC are not of great assistance in drawing these boundaries, because they tend to specify the consequences of classifying a procedure as research, rather than provide clear criteria as a basis for classification. The NH & MRC has said:

> It is not always easy, or possible, to see clearly the dividing line between procedures that may be considered to be established, as opposed to those which amount to research. The very fact that these difficulties exist is a reason for bringing such matters to the attention of ethics committees.[11]

This statement does not solve the problem, but passes it to IECs. An IEC composed as required by the NH & MRC[12] is not necessarily an expert body, and it is given no guidance at all to resolve doubtful cases. In practice the answer to the question of degree, whether or not to consult an IEC, must be made by the individual researcher involved. Although guidelines in regulatory codes will help, it is also necessary that the practitioner or researcher recognize the case as one falling within the guidelines.

Research classifications and the regulatory codes

The distinction between therapeutic and non-therapeutic research is whether or not the procedure is for the direct benefit of the subject. The general introduction to the Declaration of Helsinki recognizes the distinction as follows:

> *The purpose of biomedical research involving human subjects must be to improve diagnostic, therapeutic and prophylactic procedures and the understanding of the aetiology and pathogenesis of disease.*
>
> *In current medical practice most diagnostic, therapeutic or prophylactic procedures involve hazards. This applies a fortiori to biomedical research.*
>
> *Medical progress is based on research which ultimately must rest in part on experimentation involving human subjects.*
>
> *In the field of biomedical research a fundamental distinction must be recognized between medical research in which the aim is essentially diagnostic or therapeutic for a patient, and medical research, the essential object of which is purely scientific and without direct diagnostic or therapeutic value to the person subjected to the research. [Emphasis added.]*[13]

The declaration proceeds to state the general principles of research on humans, and then to state particular principles which apply to research in two categories under the headings: 'Medical research combined with clinical care (clinical research)', and 'Non-therapeutic biomedical research involving human subjects (non-clinical biomedical research)'. The two categories are based on both the therapeutic/non-therapeutic classification and the clinical/non-clinical context, which are not necessarily identical. If 'clinical context' merely refers to the fact that the research takes place within an existing doctor–patient relationship in terms that would be familiar to its intended medical audience, then it adds nothing.

The Declaration of Helsinki requires that biomedical research involving human subjects should occur only under medical supervision, and states that responsibility for the human subject must always rest with the doctor.[14] There is an ambiguity in the doctor's role in supervising research carried out by scientists. If the purpose of the research is not to benefit the subject, then it will be non-therapeutic research and the fact that it is medically supervised will not convert it into therapeutic research. In such a case, the Declaration of Helsinki contemplates a doctor–patient relationship which is not solely for the patient's benefit. If such a relationship is regarded as a clinical context for research, then the declaration's reliance on clinical context rather than therapeutic aim would be highly significant.

The doctor–patient relationship is an important context for research. When research is undertaken within an existing doctor–patient relationship, however, special precautions may be needed to ensure that the informed consent of the patient/subject is obtained freely and is unaffected by the patient's dependent position in that relationship. The importance of truly free and informed consent in this relationship may be one underlying reason for the persistence of the therapeutic/non-therapeutic distinction and its appearance in the declaration.

The distinction has, however, been criticized. It has been said that: 'many types of research cannot be defined as either therapeutic or non-therapeutic'[15] and that because the Declaration of Helsinki imposes limitations on who may be the subject of non-therapeutic research, adherence to the dichotomy would have the effect of preventing the carrying out of certain types of research at all. For instance, therapeutic drug trials using diseased control groups given placebos, or research involving those suffering from diseases which is designed to explore the causes of the disease rather than to treat it, could not be conducted under this dichotomy.

The real problem with the therapeutic/non-therapeutic dichotomy is that it represents a theoretical ideal which does not correspond with what actually occurs. In reality, no one has perfect information, motives are complex, and there are many other factors (such as need for external funding) that influence whether a procedure will be treated as research, subject to such requirements as IEC approval, or as therapy, in which case all the formalities are avoided.

In Australia, no distinction based on the purposes of research is expressly stated in the NH & MRC Statement on Human Experimentation. However, the preamble to the statement refers to the Declaration of Helsinki, so it should be read against that background. The only express reliance on the therapeutic/non-therapeutic dichotomy by the NH & MRC is in Supplementary Note 2 in relation to research on children, where the distinction is whether the procedure is of any direct benefit to the child.[16] Research with children as subjects apparently prompted the clearest expression of the need to protect the subject, which confirms that this may be the reason for the persistence of the dichotomy.

Once it is determined that a regulatory code applies to a research proposal, there are two main criteria by which its ethical acceptability is to be judged. First, the research must be worth doing and scientifically valid, and second, the informed consent of the subject must be obtained.[17] These criteria must be met independently: if a research project is not scientifically valid, then the mere fact that informed consent has been obtained from the subject will not make it acceptable. Scientific validity requires that the research be such that valid results can be obtained. To be worth doing, the risk of harm to the subject must be minimized and not disproportionate to the chance of benefit, and the benefit must be sufficient to justify doing the research.[18] It is necessary, therefore, to make an advance assessment of the purpose, risks, benefits and beneficiaries of research to judge its ethical acceptability under the regulatory codes. Such an assessment can be based only on indirect evidence, and it cannot be factually validated until the research is done.

In summary, the practical problem is to apply the distinctions between practice and research (validation) and between therapeutic and non-therapeutic research (aim or purpose). Validation is ultimately a question of degree. Classifying research by purpose raises its own problems: who judges the 'purpose' of research—the researcher or someone else; and how widely or narrowly can aim or purpose be stated? Research is classified as therapeutic when it is done for the direct benefit of a patient: but how direct, how significant and how likely must the therapeutic benefit be? Is it always realistic to distinguish between benefit to particular individual research subjects, and present benefits to other groups of individuals or to society in general, or even future benefits to the subject, given the contingent nature of much medical knowledge?

We now move on to consider how IVF and embryo research should be classified in light of these problems, then turn to the question of who is the subject of research.

Classifying IVF and embryo research

The problems of applying the regulatory codes are exacerbated when the area of concern is controversial. Two categories of activities relating to new reproductive technology need to be considered: current clinical IVF and research for the purpose of improving present IVF procedures or developing new techniques. While the two are related, they raise different issues.

The general regulatory codes for research have been supplemented in the area of IVF. Both the NH & MRC and the World Medical Association have issued specific statements on IVF which refer to both research and practice. In relation to embryo research, these statements differ from the research codes in going beyond procedural matters to matters of ethical principle: they state that embryo research is acceptable in some circumstances, and necessary for the further development of IVF.[19] These statements are expressly guided by the general aim of improving biomedical knowledge, but they do not go so far as to elaborate on the circumstances in which embryo research is acceptable.

Clinical application of IVF

Can IVF treatment be regarded as an established (or orthodox or validated) procedure and as part of medical practice, or should its clinical application still be regarded as experimental? If clinical IVF treatment is regarded as a therapeutic procedure, then who is the patient: the woman, the embryo, or both? If both, then how are conflicts between their interests to be resolved? If clinical IVF is part of medical practice, what variations from the standard treatment procedure should be considered as exceeding 'the flexible limits of the orthodox'[20] and, therefore, as research? In relation to the various kinds of embryo research to improve or further develop IVF, who is the subject of the research, and when, if ever, is it to be classified as therapeutic research?

It is clear that IVF treatment is now widely regarded as an established procedure. Its cost is borne by patients, health insurance funds and government, not by the NH & MRC.[21] It is, however, an established procedure with a difference, because it is still controlled in many respects by the NH & MRC code, which beyond this area applies only to medical research. It is interesting to note the contrast in the terminology used by Jansen and McCaughey in a 1982 background paper for the NH & MRC, and the terms adopted by the NH & MRC in its Supplementary Note on IVF and ET, based substantially on the background paper. Jansen and McCaughey wrote that:

> . . . IVF and ET can be a justifiable means of treating infertility, but it is premature to regard IVF and ET as an established therapeutic procedure. Accordingly all work in this field should be seen as experimental and subject to the guidelines for research on humans adopted by the NH & MRC.[22]

In its supplementary note, the NH & MRC stated that:

> In vitro fertilization (IVF) of human ova with human sperm and transfer of the early embryo to the human uterus (embryo transfer, ET) can be a justifiable means of treating infertility. While IVF and ET is an established procedure, much research remains to be done and the NH & MRC Statement on Human Experimentation should continue to apply to all work done in this field.

While Jansen and McCaughey considered that IVF could not be regarded as an established therapeutic procedure but was still experimental, the NH & MRC called IVF an established procedure, without indicating the basis for this claim: was it the existence of validation in Levine's terms, or was it the fact of social or medical acceptance of IVF as a medical technique? Both extracts state that the NH & MRC guidelines should continue to apply to 'all work in the field', but do not define this phrase—does it apply only to research work, or also to clinical application? This aspect is clarified by the Supplementary Note on IVF, which requires that 'every centre or institution offering an IVF and ET program should have all aspects of the program approved by an institutional ethics committee'.[23] In effect, clinical IVF must be done in accordance with an approved procedure, and it would follow that significant variations from the approved procedure would be regarded as experimental and require IEC approval.

The IEC which approves the procedure for clinical IVF is in the position of defining the 'established procedure'. This example clearly raises problems concerning the process of moving from therapeutic research to established practice. The problems centre on the question of validation or 'establishment': for example, is success rate relevant to validation, and if so, what success rate would IVF need to achieve before it could be regarded as clinically established and not subject to regulatory codes designed for research?[24] Is a low success rate, such as is the case now, sufficient for validation in the absence of any alternative treatment? Should success rate alone be the measure for validation, or is some other criterion such as acceptance by the medical profession also necessary? If so, how is the latter to be established—is a decision by the NH & MRC either necessary, or sufficient?

Accepting that IVF and ET is an established procedure, what activities are covered? The answer depends on the procedures approved by IECs, but there is little guidance for IECs as to how specific their approval must be. Is it possible for the individual techniques in the IVF procedure, such as *in vitro* culture of the early embryo, to be regarded as still experimental even though the treatment overall is regarded as established? Should the treatment be considered as a whole or can it be subdivided into separate elements, such as egg collection, fertilization, embryo transfer, pregnancy and birth?

Whatever the answers to these questions, some procedures will fall outside the approved protocol for clinical IVF and must then be treated as experimental.

Research to develop or improve IVF techniques

The NH & MRC Supplementary Note on IVF and World Medical Association 1984 Statement on Human IVF state the necessity of some embryo research, but go no further. Therefore we have to return to the general provisions of the regulatory codes for guidelines applicable to IVF embryo research.

As noted above, the Declaration of Helsinki, but not the NH & MRC Statement, expressly limits the purposes for which biomedical research can be done. At the present time, human embryo research could only be regarded as (loosely) therapeutic if it is designed to improve the success rate of current IVF treatment or to correct a defect in a particular embryo, although the

latter is not at present technically possible. Improvement of existing techniques needs to be distinguished from the development of new techniques
for creating embryos (such as microinjection of sperm or from frozen eggs).
In the latter cases destructive embryo research is more likely to be necessary,
which is clearly not therapeutic for the particular embryo.

For improvements in existing techniques, at least, variations of *sufficient* significance in any IVF procedures should be regarded as research and
require approval by an IEC. The limits of the treatment discretion are
particularly difficult to state with a new treatment like IVF, and the degree
of variation which will attract the requirement for IEC approval correspondingly difficult to determine. A specific example may help to clarify the
problem: when does a variation to the culture medium used to nurture the
human egg and sperm in the laboratory before and after fertilization require
ethical approval? If all IVF improvement work is experimental, this must
include variations in any individual procedures unless the variation is in-
significant. In its 1987 Review of IVF Centres, the NH & MRC stated:

> . . . *opinions differ about what amounts to research and experiment.
> We therefore enquired beyond the recorded research projects at each
> institution. It became apparent that some centres held the view that
> certain activities did not amount to research, or at least to the kind of
> research that requires the specific approval of an institutional ethics
> committee. A common activity is one which involves changing the
> ingredients which comprise the culture medium or nutrient solution in
> which fertilized ova are placed prior to transfer. Another is variation in
> the method of manipulating the fertilized ovum. Most centres held the
> view that minor changes in these procedures did not amount to research
> or experimentation or, at least, did not require specific institutional ethics
> committee approval. We consider, however, that all such changes and
> variations do amount to research because they aim to permit valid
> conclusions to be reached on whether what is being tried is beneficial . . .
> In order to be ethically acceptable, such activities must be scientifical.y
> valid as well as comply with other ethical criteria.* [25]

This statement gives the NH & MRC system broad application, consistent
with the NH & MRC's earlier quoted comments about referring to an IEC
when in doubt. Once an activity such as culture medium variation is characterized as research, the crucial question is raised: who can give informed
consent to it as required for ethical approval, and who should be regarded
as the subject of such research? Although clearly the embryo actually undergoes the variations, an embryo itself is incapable of giving informed consent.
In such a case the embryo will be transferred to a woman, who is also directly
affected by any research, and is capable of giving consent.

Under its Supplementary Note on IVF, the NH & MRC says that the
people who provided the gametes from which the embryo is created should
give informed consent to research on the embryo. [26] This is consistent with

the view that the research is done to improve therapeutic treatment of the individuals providing the gametes, particularly the woman, which is the natural perspective of the medical profession. By contrast legislative restrictions on embryo research focus exclusively on the embryo as a research subject, and provide no mechanism for reconciling the interests of the embryo and of the woman in cases where both are research subjects.

Who is the subject of IVF research?

It is clear that women are the patients in IVF treatment who give consent to treatment. In most cases, all invasive treatment is carried out on the woman, although this fact can be obscured by talking about treating 'couples' through IVF. When attention is turned to research designed to improve or develop IVF techniques, however, the woman is frequently lost sight of as the subject of research. Embryo research is argued to be justified as therapeutic research because it is necessary to improve the success rate of IVF for the benefit of women undergoing the treatment. This position is confirmed by the NH & MRC Supplementary Note on IVF as well as the recent legislation. Given social concern about reproductive technology leading eventually to a Brave New World, it would be much more difficult to justify embryo research as non-therapeutic research. But when the ethics of embryo research are discussed, usually only the fertilized egg is given consideration as the research subject. As a result, therapeutic research in IVF has come to mean research that is therapeutic for the embryo and not research that is therapeutic for the woman. Although it may be easier to focus solely on the embryo when considering what may be done with it in the laboratory when it is outside the woman's uterus, it is unrealistic to treat this situation as if it had no relationship to what happens when an embryo is transferred to a woman. Research which is therapeutic for the embryo may be non-therapeutic research for the woman to whom the embryo is transferred. For example, freezing an embryo so that it may be transferred to a woman in a later cycle may increase the chances of successful implantation, but the results of the particular freezing process may not be known. The divergence of interests should not be passed over.

Clearly, research on embryos is in a different category from research which involves only a competent patient: it is justifiable, and indeed necessary, also to consider the embryo. The woman is an adult capable of giving consent not only to medical treatment but also to being the subject of research, while the embryo is not. Before research is conducted, however, two requirements of the regulatory codes must be met: not only must informed consent be given by the subject, but also the research must be scientifically valid in the sense that valid results can be obtained and the risk of harm is minimized. If the embryo is the subject of research, how is this possible?

Consider an example of research designed to develop new techniques for IVF, such as microinjection of sperm or egg-freezing. New methods for creating embryos need to be tested for genetic safety before it is possible to say under the regulatory codes that embryos created by the method may be transferred to a woman with the minimum of risk. If the woman and not the embryo is considered as the research subject, such testing, even if destructive, would be justifiable under the Helsinki Statement on Human *In Vitro* Fertilization and the NH & MRC code, provided the harm/benefit ratio was sufficiently low. If the embryo is the subject of the research, and appropriate testing cannot be carried out, because the requirements of the codes cannot be satisfied, then women in the program are left to undergo procedures that are not able to be optimally or adequately tested and, therefore, must be regarded as experimental. Women to whom embryos created through such procedures are transferred take the risk that the embryo may be defective as a result of the lack of adequate prior research. From the viewpoint of the medical regulatory codes, and the doctor's responsibilities to the female IVF patient, transfer of such embryos is of dubious acceptability, and prior testing is desirable. Although carrying a defective embryo is not the same as having a physiological disease, its effects in terms of physiological and psychological stresses on the woman would be very similar.

From this perspective, the question is not simply whether or not embryo research should be permitted. It is whether the ethical issue of how the embryo is treated should be separated from and prevail over the question of the woman's interests and treatment. If the embryo is considered to have the same moral status as a normal human being, destructive embryo research can be ruled out, even if it is in the interests of the woman. This view overlooks the complexities we have discussed which spring from the factual dependence of the embryo on a woman for gestation, which makes the win/lose model for conflict resolution seem incongruous.

On any other view of the embryo the ethical acceptability of transferring to women embryos produced through non-validated procedures should not be ignored or implicitly tolerated—it should be properly considered. Whether it amounts to therapeutic or non-therapeutic research, the regulatory codes must be complied with. The alternatives are to cease using these techniques (even though they could become validated if proper testing were permitted) or to do the necessary embryo research to validate procedures before transferring embryos created in that way to women. If the latter course is foreclosed, there is an impasse and, according to the codes, the new techniques should not be used or developed at all.

The separation of embryo research from the woman's treatment has led to a situation where women undergoing IVF treatment continue to accept the transfer of potentially defective embryos and the possibility of miscarriage, therapeutic abortion, or giving birth to a child with congenital abnormalities, because of their commitment to the goal of having a child. It is not surprising that IVF patients are supportive of embryo research, and if research were dependent on the consent of donors of eggs or embryos, then there are strong reasons why they might consent, although there are also equally strong reasons why they would prefer to conserve their own ova for use in their own IVF treatment rather than research.[27] The regulatory codes need clearer definition of who is the 'subject' (or who are the subjects) of IVF research, and consideration of the full implications of the definition.

The fundamental problems of regulating embryo research lie in accommodating whichever moral position is adopted to the real context in which regulation takes place. This includes the dependence of embryos on women for gestation, and the empirical, uncertain nature of biomedical knowledge, which precludes the dogmatic application of standard ethical rules.

Notes

1 *Infertility (Medical Procedures) Act 1984* (Vic.), *Reproductive Technology Act 1988* (SA); see also the Human Embryo Experimentation Bill 1985 introduced into the Senate unsuccessfully by Senator Brian Harradine.

2 Beecher, H., 'Consent in clinical experimentation—myth and reality', *Journal of the American Medical Association* 195 (1966), 34–5.

3 See e.g. R. Levine, *Ethics and Regulation of Clinical Research* (Urban and Schwartzenburg, Baltimore 1981), pp. 6–8, discussing the work of the US President's Commission for the Study of Ethical Problems in Medicine and Biomedical and Behavioral Research.

4 Ibid., pp. 2–3.

5 *Report of the Committee of Inquiry into Allegations Concerning the Treatment of Cervical Cancer at National Women's Hospital and into Other Related Matters* (Judge Sylvia Cartwright, chair), (NZ Government Printing Office, 1988).

6 Ibid., Chapter 3, 'Was it a research proposal?'.

7 In some areas, for example the use of new drugs, certification mechanisms may exist, both directly and for purposes such as inclusion in the national pharmaceutical benefits scheme under s.101 of the *National Health Act 1963* and the *Health Insurance Act 1973* (Cth).

8 NH & MRC Supplementary Note 4, para. (1).

9 D. Gould, Introduction, in Paul A. Freund (ed.), *Experimentation with Human Subjects* (Allen & Unwin, London, 1969).

10 The NH & MRC includes a specific Supplementary Note on Clinical Trials (Supplementary Note 3) which applies to drug use, but only after initial testing has been done.

11 NH & MRC, '*In vitro* fertilization centres in Australia—their observance of the National Health and Medical Research Council guidelines' (1987), 4.

12 NH & MRC Supplementary Note 1: Institutional Ethics Committees (1982, revised 1985).

13 World Medical Association, 'Declaration of Helsinki', *New England Journal of Medicine* 271 (1964), 473–4.

14 Declaration of Helsinki, Basic Principles, para. 3.

15 Levine, op. cit., p. 6.

16 NH & MRC Supplementary Note 2: Research on Children, the Mentally Ill, those in Dependent Relationships or Comparable Situations (Including Unconscious Patients), paras A.(2) and (5), and D.(1) and (2).

17 NH & MRC Statement on Human Experimentation, paras 1, 8 and 10.

18 The actual requirements of this formulation are not self-evident.

19 NH & MRC Supp. Note 4 (IVF) para (5); 'Helsinki Statement on Human *in Vitro* Fertilization 1984', *Annals of the New York Academy of Sciences* 442 (1985), 571–2, para. 10.

20 Dickens, B., 'What is a medical experiment?' *Canadian Medical Association Journal* 113 (1975), 635–6.

21 Commonwealth Department of Community Services, discussion paper, 'Commonwealth Perspectives on IVF Funding' (April 1988).

22 Jansen, R. and McCaughey, J.D., 'A background paper on IVF and embryo transfer', Appendix 3 of NH & MRC, *Ethics in Medical Research* (AGPS, Canberra, 1983), pp. 31, 38.

23 NH & MRC Supplementary Note 4, para. (1).

24 IVF's relatively low 'take-home baby' rate may be the basis for the cautious approach to its use in Australia which is apparent from government legislation for licensing IVF providers.

25 NH & MRC, '*In vitro* fertilization centres in Australia—their observance of the National Health and Medical Research Council guidelines' (1987), 3–4.

26 NH & MRC, Supplementary Note 4: IVF and Embryo Transfer (1982), para. (6).

27 Rowland, R. 'Making women visible in the embryo experimentation debate', *Bioethics* I (1987), 179, 181.

11 | Is IVF research a threat to women's autonomy?

MARY ANNE WARREN

Many of the moral objections to *in vitro* fertilization and to research on human pre-embryos are based upon beliefs about the moral status of the pre-embryo itself. I shall consider a different moral objection to the experimental production and use of IVF pre-embryos, an objection based not on the supposed rights of pre-embryos, but on those of the persons from whom they originate. This objection is central to a number of feminist critiques of the new reproductive technologies. It is that women, in particular, do not freely choose to be subjects or donors in IVF and other reproductive research. Women's 'choices' about the new reproductive technologies, it is argued, are controlled by coercive social pressures, and thus are neither free nor autonomous.[1] If women cannot autonomously choose or refuse to be experimental subjects or donors to reproductive research, then their participation in such research cannot be viewed as an exercise of their medical/reproductive rights. Nor can it easily be defended as in their own best interests, since most women who serve as subjects or donors are not greatly benefited, and some are subjected to significant medical risks.

Medical professionals respond to this objection by pointing to the practice of obtaining patients' 'informed consent' for medical treatment or research. Today, at least in much of the world, physicians and researchers are required by law and/or institutional policy to obtain the consent of any person whose gametes or pre-embryos are used in medical or reproductive research. But is informed consent an adequate protection for women's autonomy in the context of IVF research? That is the question to which this chapter is addressed.

In the first section I examine some of the reasons for requiring autonomous consent, from male as well as female donors, for research on human gametes or pre-embryos. In the second section I consider some sexual asymmetries which make the presumption of autonomous consent more problematic in the case of women. In the third section I consider the concept of informed consent and argue that adherence to ordinary standards of informed consent is not enough to enable all competent patients to make autonomous decisions. Women's practical options with respect to reproductive therapy and research might be such that it would be inappropriate to speak of choice, even when informed consent is given.

But there is no reason to conclude that this must be the case. In the fourth section I argue that most women who consider IVF therapy are capable of making substantially autonomous decisions about donating oocytes or spare pre-embryos to reproductive research. Nevertheless, respect for the autonomy of these women requires more than adherence to the usual requirements for informed consent. In the last section, I suggest some ways to counter the danger that reproductive research will compromise women's medical and reproductive autonomy.

Why should donor consent be required?

To most people, it seems obvious that research on human gametes or pre-embryos should not be done without the consent of the individuals from whom these gametes or pre-embryos originated. The Warnock Report recommended that donor consent be required as 'a matter of good practice'.[2] Those who believe that IVF research violates the rights of pre-embryos will deny that the donors' consent can justify such research. In their view, that would be morally analogous to allowing parents to volunteer their children for dangerous medical experiments that are of no possible benefit to the children themselves.[3] But few would argue that IVF or pre-embryo research should proceed *without* the donors' consent, except perhaps in exceptional cases. (The donor consent requirement might reasonably be waived, for instance, to enable a spouse or family member to donate a deceased person's stored sperm, oocytes, or spare pre-embryos.)

What is the basis of the claimed right of individuals to authorize or veto any experimental uses of their *in vitro* gametes or pre-embryos? Is there, in fact, any such right? It is relatively easy to explain why competent persons should not be subjected to medical interventions without their consent: the infliction of medical treatment upon unwilling persons constitutes a physical assault. But *in vitro* gametes or pre-embryos are no longer part of the persons from whom they originated, and experiments on them pose no direct physical threat to the latter. Why, then, should their consent be required?

There are obvious pragmatic arguments for obtaining consent. Evidence that researchers were surreptitiously appropriating patients' gametes or pre-embryos for unauthorized purposes could do much to turn opinion against IVF research. But such practical considerations cannot be central to the right in question. If there is such a right, then the covert appropriation of gametes or pre-embryos would be wrong even if it caused researchers no detrimental publicity or other misfortune.

This right is sometimes treated as a property right, arising from explicit contractual agreements. Couples undergoing IVF treatment are usually given this right by law, and/or by the terms of the agreement between themselves and those providing IVF services. The couple's reproductive cells are treated as their legal property, which is temporarily held in trust by the IVF program, and which may be used only in ways the couple have authorized. But such legal or contractual arrangements cannot be the primary basis of this right either. If there is a general right to control the uses others make of one's gametes or pre-embryos, it must be a moral right, and not entirely derivative from specific laws or contracts. Otherwise, in the absence of legal or contractual prohibitions, researchers would do nothing seriously wrong in using human reproductive cells in ways the 'donors' would find unacceptable.

The physical origin of human reproductive cells may provide another basis for the right in question. Perhaps the right to control one's own body has, as an implication or corollary, the right to control the uses made of any tissues or cells derived from one's body. True, so-called 'discarded tissue' is not usually regarded as the patient's property, and should it be needed for some research project the patient's consent is not usually sought. But this may be appropriate only because most people do not care what is done with such material. Perhaps, when individuals do have strong preferences about what is done with cells or tissues derived from their bodies, those preferences ought to be respected.

This is a plausible claim, at least in many instances. A person's discarded tissues are related to that person in potentially important ways. For instance hair, like blood and other body fluids, can be analysed to detect the use of certain proscribed drugs—though often not very reliably. Tissue analyses may also detect disease or pre-disease conditions, the knowledge of which, in certain hands, could result in the loss of employment or health insurance. For these and other reasons, individuals have a legitimate interest in controlling what others do with their discarded tissues. That interest will become even more important as additional procedures for extracting information from human cells, tissues and body fluids are developed and marketed.

But gametes and pre-embryos are unlike (other) discarded tissues in one crucial respect. Not only are they a possible source of information about the individual from whom they came; they also have the potential to contribute to the development of a new human being, the genetic child of that individual. If it were possible to 'clone' human beings from the genetic material contained in any living human cell, then any such cell would have that potential. But that kind of cloning is currently impossible, and likely to remain so for a long time. In this respect, then, reproductive cells are unique.

I strongly doubt that the possible developmental future of gametes or pre-embryos is enough to establish moral rights *for them*.[4] But it is enough to give both women and men a legitimate interest in controlling the uses made of their reproductive cells. This developmental potential means that the unauthorized use of gametes and pre-embryos is objectionable, in part, because it may violate reproductive rights.

Like other moral rights, reproductive rights are social creations, attempts to create zones of protection for certain interests that are vital to individual or societal well-being. The right to reproductive autonomy protects the powerful interest that most people have in controlling their own reproductive lives—or at least in not being unjustly prevented from doing so. Reproductive autonomy is a special case of the right to self-determination, which many regard as the most basic moral right.

The right to reproductive autonomy precludes a large range of coercive interventions into the reproductive lives of individuals, particularly (but not exclusively) interventions which prevent them from having children when they wish to or which force them to have children when they wish not to. It is, to that extent, a 'negative' right, i.e. a right that others refrain from certain actions. But, like other moral rights, it also sometimes requires that positive steps be taken to enable individuals to enjoy the benefits which the right is intended to protect.

On most analyses of the right to reproductive autonomy, some coercive limitations of individuals' medical/reproductive options may be permissible. It might, for instance, be appropriate to ban the sale of some medically dangerous contraceptive—provided that safer and equally affordable substitutes are available. But reproductive autonomy requires that such exceptions be made only on the basis of a clear necessity. Those who wish coercively to limit the medical/reproductive choices of others must show, not only that there are some probable net benefits from these limitations, but also that these limitations are likely to be less objectionable and costly to each of the affected individuals than the alternative policy of letting them make their own decisions, on the basis of the best information available.

Should the right to reproductive autonomy include the right to control the uses made of one's *in vitro* gametes or pre-embryos? It might seem odd that reproductive autonomy should extend to a veto over procedures that (so far as can be predicted) will neither cause the birth of unwanted children, nor prevent the birth of wanted ones. Why, for instance, should a couple undergoing IVF treatment retain the right to control the disposition of their spare pre-embryos, once they have decided not to use them to try to start a pregnancy?

One answer is that the *in vitro* fertilization of a human ovum by a human sperm is a part of the reproductive lives of at least two persons. Even if this event will probably never lead to the birth of an infant, it is at least theoretically possible that it might, e.g. with the aid of a 'surrogate' mother. That fact gives those persons a legitimate interest in this event and its eventual outcome. They may also have moral, religious or other reasons for attaching importance to the conception and pre-embryonic development of their own potential children. Some of these reasons may be debatable; but that is not the point. In the absence of clearly compelling reasons to the contrary, they are entitled to conduct their reproductive lives as they think best. Neither scientific curiosity nor the possibility of therapeutic benefits to other persons could justify the use of coercive or deceitful means to obtain human gametes or pre-embryos for research purposes.

Some sexual asymmetries

These arguments for requiring consent for IVF research are applicable both to female and to male donors. But donating gametes or pre-embryos is a different proposition for women than for men. One difference is that it is not sperm, but oocytes, which are a scarce and limiting resource in such research. Whereas most fertile men regularly produce millions of spermatozoa, women normally produce only one mature oocyte each month. Furthermore, while sperm donation usually requires no medical intervention, 'oocyte recovery' involves the physical invasion of the woman's body and, typically, the administration of drugs or hormones to alter the functioning of her reproductive system.

In contemporary IVF programs, oocytes are collected from the surface of the ovary, either through incisions in the woman's abdominal wall—the laparoscopic method[5]—or through the use of ultrasound to guide an aspiration needle.[6] While the latter method usually does not require surgery or general anaesthesia, it is still invasive, and carries some risk of infection or damage to internal organs. In addition, the woman is usually given anti-oestrogenic drugs or gonadotrophic hormones to cause her to produce multiple mature oocytes in a single monthly cycle. The long-term effects of these drugs are not yet known. Oocytes may also be obtained from women who undergo surgery for other reasons, e.g., hysterectomy, or tubal ligation for sterilization.[7] In such cases, too, oocyte recovery may create additional risks to women.

These women are potentially vulnerable to medical exploitation. The researchers' need for oocytes and pre-embryos may create a conflict of interest that adversely affects the medical care some women receive. In the worst case, women might be coerced or deceived into undergoing unnecessary surgery, drug exposure or other potentially harmful procedures in order to provide reproductive researchers with additional opportunities for the collection of oocytes. There have already been allegations of this sort of abuse. Gena Corea reports that, according to physician and author Michelle Harrison, the decision to remove the healthy ovaries of premenopausal hysterectomy patients has sometimes been made in apparent response to requests for these organs from infertility researchers.[8]

The danger that women will be victimized by what Corea calls 'egg snatchers'[9] has little or no practical parallel in the case of men. Researchers are unlikely to engage in 'sperm-snatching', since sperm can generally be obtained without much difficulty from willing donors. If sperm donated for one purpose (e.g., contraceptive research) were used for some other purpose (e.g., artificial insemination), without the donor's consent, then his right to reproductive autonomy may well have been violated. But he would not, in addition, have been unwillingly or unwittingly subjected to potentially harmful medical interventions.

Women are also particularly vulnerable because of the predominantly male membership and hierarchy of the biomedical professions. Medicine has historically excluded women, and is still not free of patronizing and misogynist attitudes towards female patients. Thus, it is difficult for some medical professionals to respect women as autonomous decision-makers.

It is also difficult for many men to understand women's interests in the area of medical/reproductive care. Much of what women experience in that context is foreign to their experience. For not only do they lack a female reproductive system, but their perceptions are inevitably influenced by the different cultural world which they inhabit, as members of the dominant gender. Consequently, male physicians and researchers are prone to certain sorts of error. They may underestimate the severity of the side-effects that women experience from particular treatments, and they may disbelieve women's reports of their own symptoms. They may also fail to understand the effects upon women of the *absence* of certain medical/reproductive options; e.g., advice from a doctor who is also a woman; or home birth with a physician or midwife in attendance. Consequently, the forms of care offered by the predominantly male medical profession often limit women's options in ways that cannot be justified.

Realities outside the medical context can also limit women's reproductive autonomy. One feminist concern has been the influence of pronatalism. Pronatalism has been defined as 'any attitude or policy that . . . encourages reproduction, that exalts the role of parenthood'.[10] As Judith Blake has argued, most women and men everywhere are 'channelled' towards parenthood, e.g. by the influence of religion, education, public opinion, art and scientific theory.[11] The impact of pronatalist ideology upon women is especially severe, because cultural norms of femininity are more insistently linked to parenthood than are the equivalent standards of masculinity. It is women who are persistently exposed to the message that they cannot be worthy, happy or fulfilled unless they have children.

In Blake's view, the power of pronatalism is such that women cannot (be correctly said to) *choose* to have children.[12] Others have drawn a like conclusion about women's participation as patients or donors in reproductive research. For instance, Corea argues that women's consent to undergo IVF treatment is so strongly conditioned by the pronatalist culture that it must be seen as a product of social coercion rather than autonomous individual choice.[13] In her view, women's socially reinforced desire for motherhood is so powerful that it inevitably interferes with their capacity critically to evaluate therapies that purport to offer a means of fulfilling that desire.

Worse, it can make them vulnerable to a form of implicit blackmail, whereby they are expected to lend their bodies and cells to reproductive research in return for receiving the benefits of the new reproductive therapies.

If it is true that women cannot make autonomous decisions about their involvement as subjects or donors in reproductive research, then such research probably should not be done at all—or not until the unjust conditions that may control women's reproductive choices are overcome. But is it true? I will argue that, despite the unjust conditions that limit reproductive freedom, women are not incapable of making autonomous choices with respect to reproductive therapy or research. The argument requires more than an appeal to informed consent; however, it will be useful to begin with an examination of that concept.

Autonomy and informed consent

The view that mentally competent adults have the right to make autonomous decisions about their own medical treatment is relatively new to the medical profession. Before this century, the dominant model of the doctor–patient relationship was markedly paternalistic. Patients were generally expected to follow doctors' orders, not make decisions about their own therapy. What they were told about their medical condition, the nature of the proposed therapy, its risks and possible benefits, and any possible therapeutic alternatives, was largely left to the discretion of the physician.[14]

In contrast, the contemporary ideal of patient autonomy treats the doctor more as an expert advisor to the patient than as one empowered to issue orders. Since the 1950s, legal and other regulatory protections of patients' autonomy in some countries have often included the requirement that informed consent be obtained for any treatment or experimental research undertaken. In *A History and Theory of Informed Consent* Ruth Faden and Tom Beauchamp present an analysis of the concept of informed consent which captures its essential features.

Faden and Beauchamp define informed consent as *autonomous* consent. Autonomous actions are those which are performed intentionally, with understanding, and in the absence of improper controlling influences.[15] Autonomy, they note, is a practical concept. When we ask whether some action is autonomous, we are not asking whether it is fully autonomous, i.e., performed with total understanding and in total freedom from any inappropriate outside influence; for that is an impossible ideal. In practical contexts such as medicine, what is required is not full autonomy but rather substantial autonomy.

Substantial autonomy requires not total understanding, but a reasonable grasp of the clearly relevant information; and not total freedom from all improper influences, but the absence of coercion or morally objectionable forms of manipulation. Substantially autonomous decision-making by patients requires that they be provided with adequate, accurate and comprehensible information, and that their consent not be obtained through deliberate coercion or improper manipulation. Patient autonomy is the moral goal which ought to motivate the implementation of informed consent regulations. It is, they conclude, 'a reasonable and achievable goal', in medicine as in other practical contexts.[16]

One reason Faden and Beauchamp find it relatively easy to equate informed consent with autonomous decision-making is that they employ definitions of coercion and wrongful manipulation which are suspiciously narrow. Coercion is defined as 'an extreme form of influence by another person that completely controls a person's decision'.[17] This influence must be intentional; typically, it involves a threat which the threatened person cannot resist. Manipulation is also defined as an intentional influence by another person, but one that could be resisted, although it is not.[18] Unlike coercion, manipulation is not necessarily inconsistent with substantial autonomous decision-making. However, autonomy can be defeated through the manipulation of information, e.g., by lying, withholding pertinent facts, or presenting facts in a deliberately misleading way. Autonomy can also be defeated by deliberate psychological manipulations, such as playing upon fear, anxiety, or guilt in order to interfere with the person's ability to understand or deliberate effectively. Offers (e.g., of financial compensation) are often manipulative, but never coercive. They are inconsistent with autonomous decision-making only if they are unwelcome—i.e., repugnant to those to whom they are made, yet very difficult for them to refuse.[19]

This account of coercion and wrongful manipulation rules out some of the most important feminist concerns about women's reproductive autonomy. On this account, coercion and wrongful manipulation can only occur through the intentional actions of individuals. But many of the social, economic and other influences that feminists hold to be coercive are not readily reducible to the intentional actions of specific individuals. Pronatalism would certainly resist any such reduction. Thus, on the account that Faden and Beauchamp give, it is not even a logically possible obstacle to autonomous choice.

On this conceptual issue, the feminist critics are surely right. Informed consent, as Faden and Beauchamp define it, is not a guarantee of autonomy. An individual's decision may not be deliberately coerced or wrongly manipulated by any other individual, yet can be controlled by social or economic factors that preclude autonomous choice. Even welcome offers may be coercive—or part of a coercive situation—if they are welcome only because other options have been unjustly foreclosed.

Sex-selective abortion may (sometimes) be a case in point. Where this service is in great demand—e.g., parts of India and South-east Asia—women are under powerful social and economic pressure to have sons. Daughters have little earning power, and will in any case leave the family once they marry; moreover, they may need large dowries. Consequently, the birth of a daughter is often regarded as a great misfortune. Unwanted daughters, and women who do not produce sons, may be abused, neglected or abandoned. In these circumstances, a woman might find it impossible to refuse to abort a female fetus, even though the prospect offends her strongest moral convictions. If social and economic conditions have effectively blocked all her other options, then her informed consent provides only the appearance of autonomy.

Reproductive autonomy may be defeated without intentional coercion or manipulation. Sometimes coercion is inherent in a particular culture or economic system. When unjust circumstances preclude autonomous choice, it is appropriate to speak of a violation of the right to reproductive autonomy —even though it may be difficult to fix the blame on any individual.

Informed consent is not necessarily autonomous consent. It is, however, a necessary condition for it. The implementation of informed consent requirements is a landmark in the evolution of the enforceable rights of patients and research subjects. The right to make informed and uncoerced decisions about one's own medical treatment is as yet incompletely won. In too much of the world, safe means of contraception and abortion remain unavailable, or available only to the affluent. At the same time, middle-class women are often sold unproven or unnecessary medical procedures—a trend illustrated by the massive overuse of Caesarean section in the United States. It is essential to extend women's autonomy with respect to both the older reproductive technologies and the newer ones such as IVF. Although the formal requirement of informed consent cannot guarantee reproductive autonomy, it is a protection we could ill afford to lose.

Autonomous choice in an unjust world

All women are influenced by pronatalism. Meredith Michaels has described motherhood as a 'narrative' into which women are born, even if they never become mothers. Relatively few women deliberately avoid motherhood altogether, and those who do often find that resisting the call of this cultural narrative requires 'a continual act of will'.[20] Yet for many women, the range of reproductive options is wider than in the past.

Throughout recorded history, most women have had little choice about becoming mothers. Married young, with no effective means of contraception or abortion, they had babies whether they wished to or not. Such knowledge about the control of human fertility as did exist was often systematically suppressed. There have always been women who have defied pronatalist pressures, through marriage resistance, voluntary celibacy or the clandestine practice of contraception, abortion or infanticide. But the greater availability of contraception and abortion, in this century, has made childbearing more often a matter for deliberate choice by women. For women who are poor, underage or subject to legal or religious prohibitions, preventing unwanted births can still be difficult and dangerous. But now there are also countless women who take for granted their ability to limit their fertility.

This expansion of (some) women's freedom not to have children is, in Mary O'Brien's words, a 'world historic event'.[21] The further growth of reproductive freedom is possible, but by no means inevitable. Women need more universal access to contraception and abortion, as well as to prenatal, obstetric and pediatric care. The prevention and treatment of involuntary infertility needs to be taken as seriously as the prevention of unwanted pregnancies and births, but not more so. Reproductive freedom requires that economic security not be contingent upon the production of children. Conversely, economic security must be consistent with parenthood; in both socialist and capitalist societies, the structure of paid work needs to be better accommodated to the needs of childrearers.

The medical profession can claim only modest credit for past expansions of reproductive freedom, which have often been opposed by much of that profession. (For instance, nineteenth-century physicians and medical associations led successful movements to criminalize abortion in the United States and elsewhere.) In much of the world, better access to the old reproductive technologies is probably a more urgent need than the development of new ones. Nevertheless, if IVF research yields improvements in contraception or the treatment of involuntary infertility, then it may eventually contribute much to the growth of reproductive freedom.

Inevitably, the new reproductive technologies also have the potential for harmful or coercive uses. Rather than expanding reproductive freedom, they might give rise to new forms of tyranny. Even without overt coercion, the influence of social expectation, medical authority and media oversell could make it all but impossible for infertile women *not* to undergo IVF, or whatever reproductive therapy is in vogue. Just as it is now very difficult for many women to avoid unnecessary obstetric interventions, so it may become difficult to avoid dangerous and unnecessary interventions into the processes of conception and gestation. Women will need to make decisions about the use of the new reproductive technologies that are not only autonomous but also wise and clear-sighted. Otherwise, these technologies may some day be used against women, rather than by and for women.

But for now the question is whether the situation of women who undergo innovative reproductive therapies such as IVF makes autonomous choice impossible. Two facts suggest that it does not. First, pronatalist influences have by no means eliminated these women's capacity for intelligent deliberation about reproductive technologies. Many have already used reproductive technologies to postpone childbearing. Given what they have already been through in the pursuit of motherhood, they often may be somewhat more reflective about their reasons for wanting children than the average prospective parent.

Second, most of these women have been subject not only to *pro*natalist social influences, but also to powerful *anti*natalist influences. The high cost of IVF treatment ensures that it will be marketed primarily to middle-class women. As Germaine Greer has argued, there are aspects of the western middle-class lifestyle which strongly discourage childbearing.[22] Rearing children has become very expensive, and difficult for most women to combine with paid work—which is increasingly essential for economic survival. These facts help to explain the transition to lower birth rates that generally coincides with industrialization and economic development. When raising children becomes more difficult, people tend to raise fewer children, in spite of pronatalist institutions and ideologies.

It is unlikely, therefore, that the choice to undergo IVF treatment will be made unreflectively. Women are not under economic pressure to undergo IVF therapy; on the contrary. If they conclude that for them the possible benefits are worth the costs and risks, then it is reasonable to presume that their decision is substantially autonomous. They cannot prove that their autonomy has not been in some way impaired, but it is not up to them to prove this. In the absence of decisive evidence to the contrary, the presumption must be that a competent adult is capable of making autonomous decisions about her own medical treatment.

This presumption of patient autonomy is essential. Health care providers must be reluctant to disregard a patient's expressed will, without compelling evidence that this particular person is not acting autonomously in this particular instance. Any choice an individual makes about her own medical treatment may have been controlled by some subtle form of social coercion. But this possibility cannot be used to discredit that choice, or to deny her the right to make it. If it could, then medical paternalism would replace patient autonomy altogether. Unless the burden of proof is firmly placed upon those who would deny a patient's capacity to make autonomous choices, there will be unlimited scope for coercive paternalistic intervention.

Complete reproductive freedom is a utopian ideal; but partial reproductive freedom is better than none. The long-term value to women of IVF and other new reproductive technologies remains to be seen. For that very reason, it is vital that individual women's decisions about the use of IVF be respected. Neither physicians nor legislators have the wisdom to override women's own informed judgements about matters so central to their reproductive lives.

If these reflections are correct, then IVF and pre-embryo research need not violate patients' or donors' rights to medical and reproductive autonomy. Yet obtaining patients' informed consent does not exhaust the obligation of therapists and researchers to respect patients' autonomy. They also need to be alert to the possibility that some patients are subject to covert forms of coercion or manipulation. And, as I shall argue below, they need to avoid inadvertently exploiting the unjust circumstances to which patients may be subject.

Beyond informed consent

There are circumstances in which proceeding on the basis of informed consent may not be the best way to show respect for autonomy. Sometimes there is a fine line between helping individuals whose reproductive options have already been unjustly limited to carry out the best choice that is now possible, and complicity with the forces that have created those unjust limitations. There are, however, some dangers which reproductive researchers need to recognize.

First, they should be extremely reluctant to purchase the cooperation of subjects and donors. Modest financial compensation for minor inconveniences may sometimes be appropriate, as in the case of sperm donation. But it is morally problematic for medical researchers to use financial incentives to induce individuals to undergo significant and largely unpredictable risks. Oocyte collection or embryo donation can subject women to substantial risks to health, fertility, and sometimes life itself. To recruit paid volunteers is to risk exploiting vulnerable individuals, whose consent is based upon desperation.

But why should poor or financially pressed individuals not be free to sell their services as subjects or cell donors? Would it not be unfair to deprive them of what might be much-needed income? Some women might be quite willing to sell their own oocytes or pre-embryos, and might benefit from the transaction. But others might agree because of extreme financial need, and despite powerful misgivings. For them, the offer might be unwelcome, but difficult to refuse. In these instances, the transaction would be exploitative, and possibly a violation of reproductive autonomy. It is the exploitation, not the sale itself, which is wrong. But the pervasiveness of what has been called 'the feminization of poverty' suggests that exploitation will be difficult to avoid if women are offered substantial sums for serving as subjects or donors to IVF research.

A second problem arises from the relative powerlessness of those who may be asked to donate reproductive cells for research purposes. Asking individuals to participate in research that involves substantial risk to themselves is morally problematic when those individuals are in a powerless and dependent situation: if, for instance, they are prisoners. Even if no reward or penalty is announced in connection with the request, powerless individuals may rationally believe that they must avoid the displeasure of those who control their fate. They may, therefore, agree because of the fear of reprisals if they do not.

While the circumstances of IVF patients are not usually comparable to those of prisoners, nevertheless they often perceive their situation as one of dependence and powerlessness. They may fear that, should they decline to cooperate with any research projects associated with the IVF program, they may be dropped from the program, or otherwise penalized. Thus, it should consistently (and truthfully) be stressed that they will under no circumstances be penalized, either directly or indirectly, should they decide not to donate gametes or pre-embryos for research purposes. Unless this is mutually understood, their consent may be obtained through the unintentional exploitation of their fear of the consequences of refusal.

A third danger is the creation, encouragement or exploitation of unrealistic hope. There are, by many accounts, women who seek IVF therapy or have it recommended to them, but whose chances of success are clearly remote. IVF applicants have sometimes objected to the use of selection criteria, such as age, to exclude from IVF programs women whose chances of a successful pregnancy are thought to be unacceptably low.[23] But while any particular criterion may be subject to debate, there must be some point beyond which the odds of success are so slight that it would be wrong to provide IVF treatments on the basis of so small a hope. If women in this situation are offered IVF therapy, and also asked to donate reproductive cells for research, then both therapists and researchers may be exploiting this unreasonable hope.

These are some of the dangers that reproductive researchers need to avoid if they are to respect the autonomy of women, as subjects and donors. I do not suggest that these dangers are more severe in reproductive research than in other medical research. The point is, rather, that respect for autonomy cannot be construed strictly in terms of informed consent. Equally essential is an awareness of the economic, social and psychological pressures faced by potential subjects or donors, and of the ways that reproductive research might wrongly exploit those pressures.

Is it realistic to expect such an awareness from medical professionals who, after all, are not usually social theorists or psychologists? It will certainly call for better communication than has often occurred between patients and health care providers. It will require not just the communication of information from the professional to the patient, but attentive, responsive dialogue, aimed at bringing to light any coercive or exploitative situation that might exist. Mandatory counselling is resented by some IVF patients as paternalistic and an invasion of privacy. However, given the potential for conflicts between the interests of patients and those of physicians and researchers, it is important that independent professional counselling be available on a voluntary basis.

Efforts should also be made to extend the protection offered by standard procedures for obtaining informed consent, by incorporating more explicit safeguards against some of the forms of covert coercion or exploitation that are of concern to feminists. If such efforts are successful, then informed consent requirements may come to provide a more nearly adequate protection for medical and reproductive autonomy.

The goal of reproductive research must be not only to respect the autonomy of individual patients and donors, but also to expand reproductive freedom in the long run. To this end, there must be more equal participation in the medical and research professions by women and members of racial and other minority groups. Researchers must understand the effects of what they do, not just on cells and organisms, but also on human lives. To do that, they will need the insights of those who are personally affected by the new reproductive technologies.

Summary and conclusion

Research on IVF pre-embryos, if carried out without the autonomous consent of the donors, would violate their right to reproductive autonomy. In the case of female donors, it might also subject them to substantial medical dangers. Informed consent cannot guarantee that autonomous choice will be possible, because coercive circumstances can render women's choices about reproductive research less than substantially autonomous.

But it does not follow that such research should not be done. Autonomous decision-making does not presuppose an improbable freedom from all unjust or inappropriate social influences. On the contrary, respect for autonomy requires the presumption that most people can make substantially autonomous decisions about their own reproductive lives and research participation—provided that they are given an opportunity to do so. Women who seek IVF therapy are probably at least as capable of making autonomous choices about their reproductive lives as are most persons of their time and culture. Informed consent requirements cannot ensure autonomous choice, but they can reduce the risk that women (or men) will be coerced, deceived or underinformed by reproductive therapists or researchers.

Reproductive therapists and researchers are not obliged to ascertain that patients' decisions are entirely wise or ultimately free, before accepting them as autonomous. Yet they need to be alert to the possibility that certain patients' or donors' consent is controlled by unjust circumstances. In such cases, they must ask whether to take that consent at face value is to respect the autonomy of those individuals, or might instead constitute a form of complicity with those injustices. They must also be sensitive to the inadvertent exploitation of economic need, individual powerlessness or unreasonable hope. These moral pitfalls take on a special urgency in the context of IVF and pre-embryo research, where both women's reproductive autonomy and their physical integrity may be at stake.

Notes

1 See Gena Corea, *The Mother Machine: Reproductive Technologies from Artificial Insemination to Artificial Wombs* (Harper & Row, New York, 1985), p. 166, and

Jalna Hammer, 'A womb of one's own' in Rita Arditti, Renate Duelli Klein and Shelly Minden (eds), *Test-tube Women: What Future for Motherhood?* (Pandora Press, London, 1984), pp. 438–48.

2 Warnock, Mary, *A Question of Life: The Warnock Report on Human Fertilization and Embryology* (Basil Blackwell, New York, 1984), p. 81.

3 See Paul Ramsey, *The Ethics of Fetal Research* (Yale University Press, New Haven, Conn., 1975), p. 28.

4 See M. Warren, 'Do potential people have moral rights?', *Canadian Journal of Philosophy* 7, 2 (1977).

5 Webster, J., 'Laparoscopic oocyte recovery' in S. Fishel and E. M. Symonds (eds), *In Vitro Fertilization: Past, Present, Future* (IRL Press, Oxford, 1986), pp. 69–76.

6 Wikland, Matts, Hamberger, Lars and Enk, Lennart, 'Ultrasound for oocyte recovery' in Fishel and Symonds, op. cit. (see note 5 above), pp. 59–67.

7 McLaren, A., 'Discussion' in CIBA Foundation, *Human Embryo Research: Yes or No?* (Tavistock, London, 1986), p. 203.

8 Corea, Gena, 'Egg snatchers' in Arditte, Klein and Minden, op. cit. (see note 1 above), p. 39.

9 Ibid.

10 Peck, Ellen and Senderowitz, Judith (eds), *Pronatalism: The Myth of Mom and Apple Pie* (Thomas J. Crowell, New York, 1974), p. 1.

11 Blake, Judith, 'Coercive pronatalism and the American population policy', in Peck and Senderowitz, op. cit., pp. 26–97.

12 Ibid., p. 66.

13 Corea, *The Mother Machine*, p. 169.

14 See, for instance, Jay Katz, *The Silent World of Doctor and Patient* (The Free Press, New York, 1984).

15 Faden, R. and Beauchamp, T., *A History and Theory of Informed Consent* (Oxford University Press, Oxford, 1986), p. 238.

16 Ibid., p. 241.

17 Ibid., p. 330.

18 Ibid., p. 354.

19 Ibid., p. 359.

20 Michaels, Meredith W., 'Contraception, freedom and destiny: a womb of one's own' in Stuart F. Spicker, William B. Bondeson and H. Tristram Engelhardt, Jr (eds), *The Contraceptive Ethos* (D. Reidel Company, Dordrecht, Boston, 1987), p. 218.

21 O'Brien, Mary, *The Politics of Reproduction* (Routledge & Kegan Paul, London, 1981), p. 21.

22 Greer, Germaine, *Sex and Destiny: The Politics of Human Fertility* (Secker & Warburg, London, 1984).

23 See Patricia F. Brown, 'In vitro: client rights and the simple case', paper presented at the National Conference on Reproductive Technologies, sponsored by the National Feminist Network on Reproductive Technology, Canberra, May 1986.

PART 3

CONTROLLING EMBRYO EXPERIMENTATION IN A DEMOCRATIC SOCIETY

12 | Public policy and law: Possibilities and limitations

BETH GAZE

When government control or regulation of any area of activity is under consideration, the first task is to determine public policy for the area. A decision must be taken about what is to happen: should the activity be left untouched, or should the government intervene to regulate? If intervention is chosen, the aim of regulation must be identified. If the policy end to be achieved is not clearly formulated, the chance of effective regulation will be small. In a democracy the decision about intervention and the policy to be pursued must be determined through the democratic process, ultimately by the legislature, acting with advice from various sources. Once the policy is decided, then the means of implementing it have to be considered: should legislation or some other form of regulation be used, and if legislation is chosen, what form should it take? Underlying the question of what form of 'direct' regulation to use is the issue of 'indirect' control through allocation of public funds. The allocation of public resources is clearly very important, but is usually subject to much less public scrutiny and criticism than direct forms of regulation because of the inherent difficulties of comparing dissimilar claims for public support. The essays in this section discuss various aspects of the problem of formulating and implementing what I have called 'direct' policy in relation to embryo research.

On the question of embryo research, a clear division exists within the ethical field between those who take human life in any form to be sacred and those who follow a consequentialist or utilitarian approach to the question whether it is justifiable to experiment upon human embryos. It is unlikely that any public policy would be able to satisfy both these groups, let alone the many other interest groups which hold strong opinions on matters relating to reproductive technologies. Many (but not all) scientists hold the view that scientific progress is either a value-free search for knowledge, or a primary value to society, as it is in their own lives, and that it should not be impeded on grounds which appear to them to be based on emotional or illogical premises which they do not share. Many (but not all) infertile people strongly support *in vitro* fertilization and embryo research as essential to improve the medical techniques which could enable them to have children. Many (but not all) feminists object to reproductive technology on the ground that it ultimately diminishes women's autonomy and control of their reproductive capacity, and continue to debate the extent to which the technology can be separated from its social context of patriarchal control.

Conflicts of opinion and interest are not a new problem for democracies to resolve, but the debate over embryo experimentation touches fundamental issues about the sort of society in which we live and which we create for future generations. How are the conflicts to be dealt with in formulating policy towards embryo experimentation? Deciding policy by public opinion poll is not a satisfactory solution. There may be bias in the formulation of questions, and there is no guarantee that the views obtained are informed or considered opinions. In the human rights field we have long objected to the 'tyranny of the majority', and the objection is equally applicable to the ethics of embryo research. But how then is policy to be decided? Alternative approaches and methods must be developed. The chapters in PART 3 consider, from two perspectives, some of the problems of regulating. First, some of the arguments in favour of different types of policies, such as self-regulation or legislative regulation, and the problem of how to determine a public policy, are examined. Then the focus moves to implementing policy by law, and the problems of legislating.

Max Charlesworth puts the case for community control on the basis that the community has continuing responsibilities arising from IVF and its related research, and that this justifies intervention on behalf of the community to control its development. It is reasonable to expect that the law might provide a source of principles which could form a basis for, or give direction to, the development of public policy. In the first of his two contributions, Pascal Kasimba surveys the common law inherited by Australia and the United States from England for principles which could provide guidance in answering the question often asked: when, in law, does human life begin? He concludes that ultimately this approach is not very helpful in dealing with the specific problems of embryo research.

Non-intervention, or continuation of the *status quo*, should be one of the options to be considered when an area is reviewed for regulation. The absence of external regulation does not necessarily mean a field is left unregulated. Some degree of self-regulation is likely to operate in any field of activity, even if it is merely as a result of informal peer pressure. The extent of the limitations imposed by self-regulation will depend on whether it is formal or informal, and on the content of any guidelines. Australia has a formal scheme of self-regulation of medical research by guidelines and institutional ethics committees, which applies to embryo research. The adequacy of this scheme in form and in operation is clearly relevant to the question whether to intervene, and the debate between legislation and self-regulation. Pascal Kasimba examines this scheme of self-regulation and its operation in the field of IVF and related research, identifying some problems which might lead us to accept that legislative intervention may be preferable.

If we accept that legislative intervention is justifiable, the question is what policy should be followed and, in view of disagreements about IVF and embryo research, how to decide a policy. In his chapter Professor Richard Hare discusses different approaches to legislative implementation of moral principles, arguing that the concern of legislators in choosing any particular form of regulation must be with the consequences of their actions for other people who are affected by it.

When a policy direction has been chosen, there remains the problem of implementation. Legal regulation is usually done by legislation. In a sense, legal regulation will exist whether or not legislative steps are taken, because the general body of law applies to reproductive technology and embryo research in the same way as it applies to all other areas of human activity. This means that liability could exist for injuries or damage suffered by a patient or gamete donor who suffers physical or mental injury through negligence, as in the American case of *Del Zio v. The Presbyterian Hospital*.[1] But common law rules are laid down after the event, so common law is not an appropriate method for regulation in areas where a clear advance guide for action is needed. Cases are decided by judges who have no particular expertise to decide technical scientific questions, and no particular authority to decide questions of moral controversy. It is often preferable for the legislature to decide and implement a policy which can provide clear and consistent guidelines, whether by laying down rules for courts to enforce, or conferring decision- or rule-making power on a more appropriate body. For legislative regulation to be effective clear guidelines must be identified which can be incorporated in the statute, but the attempt to set clear guidelines faces problems. One of the objections often raised to permitting any research on human embryos or pre-embryos is that, because we are dealing with continuous biological processes, there is no 'marker event' of moral significance which can be used to set a time limit on research. Stephen Buckle considers this objection, arguing that lines can reasonably be drawn for pragmatic purposes even where no 'natural' marker event exists.

In 1984 the Victorian parliament enacted legislation to regulate *in vitro* fertilization and embryo research. The *Infertility (Medical Procedures) Act 1984* was the first—and is still one of the few—pieces of legislation in the world to regulate *in vitro* fertilization practice and embryo research. The legislation was based on the reports of a committee appointed by the state government to review IVF in Victoria, in response to the success of Victoria's IVF clinicians and researchers. The reports of the Committee to Consider the Social, Ethical and Legal Issues Arising from *In Vitro* Fertilization, their contribution to the development of the legislation, and the major features of the legislation relating to IVF research are discussed by Beth Gaze and Pascal Kasimba. Problems in the operation of the legislation have arisen because of the Act's terminology. Many important terms, such as 'fertilization', 'embryo' and 'infertility' are not defined in the Act. The problem of whether fertilization should be seen as a process or an event was raised by researchers in Victoria seeking permission to conduct research under the Act, and prompted an amendment to make clear that research prior to the completion of fertilization at syngamy may be acceptable in Victoria. The arguments concerning the significance of syngamy are evaluated by Stephen Buckle, Karen Dawson and Peter Singer in their joint essay.

Although the discussion of the Victorian legislation has limited direct relevance outside Australia, its indirect relevance is significant. The scientific and medical field subject to regulation is truly international, and problems such as how to define the fundamental terms will have to be resolved in any country which attempts to introduce workable regulation of the area. The Victorian experience of legislation in this complex and controversial area should provide some useful food for thought in other countries.

Note

1 US District Court, SDNY 74 Civ. 3588 (CES), reprinted in *Bioethics Reporter 7* (1985). In this case damages were awarded for emotional distress caused to the plaintiffs when hospital employees destroyed their *in vitro* embryo.

13 | Community control of IVF and embryo experimentation

MAX CHARLESWORTH

Everyone recognizes that the new reproductive technologies have profound implications for the whole community and that the community has a right to exercise some kind of surveillance and control over what takes place in this new and volatile area of medical research and practice. There are, however, large and deep differences about how that control should best be exercised.

Why should there be any kind of legislative regulation or control of IVF and related research? Why not leave the control of the new reproductive technologies in the hands of the medical scientists and the scientific and medical societies? Robert Edwards, the father of IVF, for example, has opposed the idea of legislation with respect to the research and clinical aspects of human embryology. 'It would', he argues, 'be far better to leave standards of practice primarily to the scientific and medical societies . . . In this way, new procedures can be adapted to the public need, and unacceptable methods can be suppressed'.[1] Sir Gustav Nossal also objects to legal controls in this area: 'We must', he says, 'use the soft-edged, untidy, polyvalent methods of a free and decent society'. In essence, Nossal argues, apart from the protection of the common law, we should leave the regulation of biotechnology to the scientific community itself with its own network of safeguards and controls and guidelines.[2]

On the other hand, it has been argued by almost all the commissions of enquiry into reproductive technology, both in Australia and overseas, that IVF and related research clearly raises such important issues of moral, social and legal concern that it must be subject to some kind of community regulation. The report of the Council of Europe's Ad Hoc Committee of Experts on Progress in the Biomedical Sciences is typical in this respect.

> *All these techniques [of human artificial procreation] raise problems of*
> *public health, for instance qualifications of medical teams, provision of the*
> *necessary means, equality of access to medical services in the framework*
> *of a right to health care. Furthermore, legal problems concerning relations*
> *between persons are posed, in particular in the field of parental*
> *responsibilities, right to secrecy, medical liability. The development of*
> *these techniques, their increasing use in society, the procedures involving*
> *embryos which are related to these techniques, have also raised problems*
> *of human rights already recognized and of new possible rights whose*
> *protection is claimed. Over and above all these problems, a fundamental*
> *ethical question is raised, that is to say, what attitude must man take in*
> *respect of these new powers over his destiny that science has given him.*
> *On this question, public opinion is divided in differing positions which*
> *range from an extremely liberal attitude to a very restrictive interpretation*
> *of the applications of science in this field. Doctors and scientists, for their*
> *part, have started to raise questions of their new responsibilities.*[3]

However, as we know, even where there is some agreement on the need
for regulation of IVF research, there is wide disagreement as to what aspects
of reproductive science require regulation. A survey by Jean Cohen, the
chairman of the ethics committee of the European Society of Human Reproduction and Embryology, has shown the large degree of disagreement
even between researchers and clinicians in the field of IVF research.[4] Thus,
for instance, while most European researchers agree that experimentation
of some kind on human embryos is scientifically necessary, there is disagreement over whether embryos should be created specifically for experimentation or whether only 'spares' left over from IVF procedures should be
used. Again some insist that there should be 'parental consent' for experimentation on embryos, but Robert Edwards argues that 'while informed
consent is a nice idea . . . in practice it is difficult to achieve. It is difficult
to explain to a patient what we want to do to an embryo'.[5] Further, some
of the respondents to Cohen's survey argue that there should be no absolute
limit to the time human embryos should be available for experimentation,
while others have suggested a limit of 14 days (the beginning of the embryo's
nervous system) and a French embryologist has, with Gallic logic, suggested
12 weeks since this is the limit for legal abortion in France.

On the question of legal protection for embryos, some of Cohen's
scientists doubt whether embryos need to be protected by the law, any more
than sperm and ova are at the moment. Others acknowledge the need for
some kind of legal protection against the commercial exploitation of embryos, though not the full legal protection afforded to newborn children.
Others consider that parental rights over embryos need to be established
through legislation, and an English respondent argues, with true British
phlegm, that the embryo should be seen as a chattel owned jointly by husband
and wife and able to be disposed of as they see fit. The same disagreement
obtains over who should have access to IVF. While apparently most clinics

restrict IVF to heterosexual couples, either married or in a stable relationship, some have raised doubts as to whether medical scientists can or should screen out 'unsuitable' parents. Edwards, for example, does not offer IVF to either lesbian couples or single women, but he has remarked that in the case of IVF being used with a frozen embryo thawed out after the husband's death, he would be prepared to go along with it. As he has put it: 'At least then the embryo would be conceived in love, albeit in a laboratory, and I should be prepared to be overruled in such cases'.[6]

From quite another point of view, some women's groups have quite properly complained that most discussions so far have not considered the welfare of the women involved. Thus the New South Wales Women's Advisory Council's submission has argued that the 'dignity, interests, health and positive welfare of women' must be taken into account when we are considering the control of research in reproductive technology.[7]

There is then disagreement about (1) whether *any* specific control or regulation is needed in the field of IVF and related research, and (2) among those who admit the need for regulation, as to *what aspects* of IVF research needs regulation. In such a situation, it seems to me, there is a great danger that we will end up with *ad hoc*, knee-jerk, short-term control and regulation unless we are able to formulate some general principles governing society's intervention in such areas. We must, in fact, go back to the general and fundamental question: on what grounds is society, either through formal legal means or other means, justified in controlling or regulating the activities of its members? Or, to make that question more specific, in a liberal democratic society, characterized by moral and religious pluralism, when may the State legitimately step in to regulate the activities of its members? Unless we have some general theory about the relationship between law and morality we will be forced by popular feeling, or sectional pressures, or political expediency into inappropriate legislation.

In a liberal democratic and pluralist society the relationship between morality and the law is very complex in that while there is, of course, a connection between the two realms, they do not coincide. The law is not really concerned with the enforcement of morality but rather with providing a framework of peace and order within which people may exercise their personal liberty to the greatest possible extent and make their own personal moral choices and engage in what John Stuart Mill calls their own 'experiments in living'. In a liberal democratic society I may believe as I like and do as I like, provided that I do not, by the expression of my beliefs or by my actions, cause 'harm' to others and prevent them from exercising their liberty to do as they please. We cannot then expect the law to seek to enforce virtue in general and prohibit vice in general; rather the law restricts itself to a very small area of immorality—those activities that harm others and prevent them from exercising their own free choice in matters of morality. An important part of this function of the law is the protection of the rights of those who are unable to protect their own rights.

We must, then, not expect the law to be the agency by which a common morality should be enforced in the community. Equally, we must resist the idea that if the law is silent on a particular issue, then it is condoning a line of action or conniving in it. As an American scholar has put it:

> We are a pragmatic and litigious people for whom the law is the answer to all problems, the only answer and a fully adequate answer. Thus many people confuse morality and public policy. If something is removed from the penal code, it is viewed as morally right and permissible. And if an act is seen as morally wrong, many want it made illegal. Behold the 'there ought to be a law' syndrome. This is not only conceptually wrong, it is also conversationally mischievous. It gets people with strong moral convictions locked into debates about public policy, as if only one public policy were possible given a certain moral position.[8]

It may then very well be the case that some of the practices and procedures in the area of biotechnology are held to be immoral or unethical by many people, but that nevertheless they are not made illegal or subject to legal control. To be made illegal it has to be shown not just that they are unethical or raise social problems but in addition that they are likely to have harmful implications for others, i.e. violating people's rights in some clear and obvious way. It needs also to be shown that the prohibitions of the law are likely to be obeyed by the generality of people and that enforcement of the law will not bring about more harm than good in a society where there is a plurality of widely differering moral views and convictions. One may be uneasy about the moral implications of certain aspects of biotechnology, while nevertheless accepting the fact that formal legal control would be inappropriate or counterproductive.

This disjunction between the sphere of law and the sphere of morality cuts both ways. For if to settle the moral question about IVF is not *eo ipso* to settle the legal question as to whether and how IVF should be controlled and regulated, so also to settle the legal question is not *eo ipso* to settle the moral question. In other words, we must resist the idea that if the law is silent about a given area then 'anything goes' in that area. Some bioscientists say things such as: 'I'll press ahead with my research until the law tells me to stop', as though the silence of the law enables them to escape making any ethical judgements of their own on the work they are doing. None of us can lead our normal lives simply by acting within the law; so also bioscientists cannot really fulfil their responsibilities as scientists and as human beings simply by acting 'within the law' and pressing ahead with their experiments until the law calls a halt.

What then in general terms is the proper province of specifically legal control and regulation with regard to the new reproductive technology? Apart from the important work of legal tidying-up that is necessary to clarify the status and the rights of those—parents, donors, surrogates, children, embryos, gametes—involved in IVF, I would argue that a fundamental reason for legal control and regulation is the protection of the rights of the children, and the rights of the embryo and fetus, brought into being by IVF. This is what the Report of the Family Law Council of Australia, *Creating Children*, calls 'the paramountcy of the welfare and interests of the child born of reproductive technology'.[9] I believe that this perspective is the right one, namely that IVF should be seen primarily as a way of creating children and, if one may so put it, only incidentally as a means of alleviating infertility in couples or as an area of medical research. It is that fundamentally which provides justification for society stepping in and setting up some kind of specifically legal regulation and control. As I said before, there may well be other aspects of IVF that may cause ethical unease, but this is of itself not enough to justify society's intervention, at least in a liberal democratic society where it is not the function of the law to concern itself directly with moral issues as such.

It has been argued that in essence there is no difference between *in vivo* conception and *in vitro* conception, and that just as it would be foolish to attempt to regulate *in vivo* conception by law (as it has been put, the law has no place in the bedroom) so also with *in vitro* conception. Further, we no longer see abortion as requiring legal control, so why should we think that the destruction of *in vitro* embryos or experimentation upon them should be legally regulated? Once again, we may have strong moral views against both abortion and embryo experimentation, but it does not follow that there should be legal control of either.

However, in my view, there is a crucial material difference between the *in vivo* embryo and the *in vitro* embryo. IVF involves separating conception from sexual intercourse and the *in vitro* embryo has in a sense an existence of its own and is, so to speak, in a particularly exposed position as compared with the *in vivo* embryo. It may be frozen and stored; it may be implanted in a surrogate mother; it may be experimented upon; it may be used as a source of tissue; it may even be bought and sold. I do not wish to claim that the embryo has exactly the same kind of 'rights' as the child, but it does have some *prima facie* rights that, given its peculiar situation and the real possibility of its exploitation, need protection.

There is an analogy, I would argue, between the *in vitro* embryo and the adopted child. Adoption is subject to stringent legal control in our society, and is not simply an unregulated business where people may do as they like, because the child to be adopted is seen as having interests and rights and as being in a relatively exposed and defenceless situation. Adoption laws are put in place to protect the rights of the child and to secure 'the best interests of the child' and one could, perhaps, see the legal regulation of IVF as having very much the same purpose, though in the case of the *in vitro* embryo there may be circumstances where the *prima facie* rights of the embryo may be overridden.

While, however, regulation has an appropriate place in the regulation of IVF, we must also admit that the law is a crude instrument and needs to be complemented by alternative forms of control. It is not just an either/or choice between so-called self regulation by medical scientists and formal legal regulation. The National Consultative Ethics Committee for the Life and Health Sciences, established by President Mitterrand in France in 1983, is an example of an alternative mode of regulation in this new and volatile field. As Mr Justice Kirby has put it:

> *Unless we can develop institutions that help the democratic arm of government to offer solutions to bioethical questions, it will continue to fall to the unelected judiciary (and to a lesser extent the unelected bureaucracy) to weigh the public policies involved and to provide the answers. The court-room is a good venue for the resolution of factual disputes between parties, where the issues are narrowly focused. It is an imperfect venue for the resolution of large philosophical quandaries, based on ill-understood technological developments and restricted to the parties and their lawyers—with little or no help from philosophers, theologians and the community.* [10]

Notes

1 Edwards, R.G., 'Test-tube babies: the ethical debate', *The Listener*, 27 October 1983, 19.

2 Nossal, G.J.V., 'The impact of genetic engineering on modern medicine', The First Ian McLennan Oration, University of Melbourne, 1983, p. 14.

3 Council of Europe Ad Hoc Committee of Experts in the Biomedical Sciences, 'Provisional principles on the techniques of human artificial procreation and certain procedures carried out in embryos in connection with those techniques', Strasbourg, 5 March 1985, p. 10.

4 The following details are taken from the report on the survey by Gail Vines, 'Whose baby is it anyway?', *New Scientist*, 3 July 1986, 26–7.

5 Edwards, op. cit., 20.

6 Ibid.

7 Proposal for a National Commission on Reproductive Technologies, unpublished submission by the New South Wales Women's Advisory Council to the New South Wales Government, 23 July 1986.

8 McCormick, Richard A., *How Brave a New World? Dilemmas in Bioethics* (Doubleday, New York, 1981), p. 185.

9 Family Law Council, *Creating Children: A Uniform Approach to the Law and Practice of Reproductive Technologies in Australia* (AGPS, Canberra, 1985).

10 Foreword to Alan Nichols and Trevor Hogan, *Making Babies: The Test Tube and Christian Ethics* (Acorn Press, Canberra, 1984), p. 14.

14 | IVF regulation: The search for a legal basis

PASCAL KASIMBA

In the legal field, the dawn of embryo experimentation seems to have been unexpected. Following early IVF successes, there has been a flurry of government-appointed inquiries in several countries.[1] However, not many countries have followed up with legislation or other form of regulation; instead, the debate on how to regulate embryo research continues.

This chapter addresses the legal status of the embryo in Anglo-common law in both its criminal and civil aspects in an attempt to find a legal basis on which coherent regulation could be predicated. A word on the terminology used in relation to the embryo is pertinent at this stage. In the legislation and cases that will be dealt with below, the term 'embryo' is usually used interchangeably with 'fetus', 'prenate' or even 'unborn child'. The term has, however, a clear meaning in biology and describes the development of a fertilized ovum from fertilization to about eight weeks.[2] Further, unless the context indicates otherwise, the term 'embryo' will refer to both *in vitro* and *in vivo* embryos.

The beginning of human life: A legal enigma

Although the question of the beginning of human life does not sound like a legal matter and should preferably concern philosophers, biologists and theologians rather than lawyers, it is nonetheless central to the whole idea of law in most legal systems. In the common law system with which we are here concerned the question has been of the utmost relevance for both the criminal and civil law. In criminal law, for instance, the relevance of the question of the beginning of human life is exhibited in the need to define 'person' for purposes of the category of 'offences against the person'.[3] In the area of civil law the gravity of the question is evident in the fields of succession law, marriage and domestic relations and the rapidly burgeoning law of negligence in the area of recovery for prenatal injuries.

Yet despite all the above, in law an answer to the question of the beginning of life has been very elusive. The courts have also confused the question by scattered dicta, some of it disclaiming capacity to answer it and some purportedly answering it. In the 1973 landmark decision of *Roe v. Wade*, where the United States Supreme Court was dealing with the heated issue of abortion, the court made the classic disclaimer:

> *We need not resolve the difficult question of when life begins. When those trained in the respective disciplines of medicine, philosophy, and theology are unable to arrive at any consensus the judiciary, at this point in the development of man's knowledge, is not in a position to speculate as to the answer.*[4]

In England similar reluctance is evidenced, for instance, by *Paton v. British Pregnancy Advisory Service Trustees* (1979) where Sir George Baker P stated:

> *In the discussion of human affairs and especially of abortion, controversy can rage over the moral rights, duties, interests, standards and religious views of the parties. Moral values are in issue. I am, in fact, concerned with none of these matters. I am concerned and concerned only with the law of England as it applies to this claim. My task is to apply the law free of emotion or predilection.*[5]

In other cases, however, the courts have not viewed their role as so restricted. In *Renslow v. Mennonite Hospital* (1977) the Supreme Court of Illinois went so far as to say:

> *As medical science progressed, the courts took notice that a fetus is a separate human entity prior to birth. It is by now commonly accepted that at conception the egg and sperm unite to jointly provide the genetic material for human life. Thus various courts have gradually come to recognise that the embryo from the moment of conception is a separate organism that can be compensated for negligently inflicted harm.*[6]

In Australia, similar dicta can, for example, be found in *Watt v. Rama* (1972) by Gillard J that:

> *without claiming any pretensions of being authoritative on the matter, from reading the vast material placed before us . . . it became apparent that biologically a person's well-being can be influenced, both universally and beneficially, by its pre-natal history and experience. Disease and trauma happening at any time from the womb to the tomb apparently can affect one's well-being and future health. It is obvious that 'the person'*

> who is conceived and develops in the mother's body is biologically the
> same 'person' who survives birth, lives and finally dies. [7]

These statements do not, however, help in the determination of the beginning of life. They are contradictory, and those which place the beginning
of life at conception were *obiter* and extravagant in the context of the cases
in which they were uttered, as they were not relied upon in reaching decisions. However, they illustrate the confused state of the law on the question
of the beginning of human life.

The law is more definite, though eclectic, in its attempts to answer
the recurring question of the broader aspect of prenatal life.

The concern with prenatal life

As earlier stated, both criminal and civil law have in various aspects touched
on prenatal life. However, attempts to extract any coherent principles on
the legal status of the embryo/fetus are thwarted by the law's differing and
shifting standards towards prenatal life not only between the criminal law
and civil law but also sometimes within each division. [8]

In criminal law, for instance, typical provisions relate to the protection
of unborn 'children' from abortion and destruction. [9] The offence of abortion,
like that of child destruction which it predates and to which it is related,
developed essentially to protect prenatal life against termination. This
rationale is restated in *R v. Bourne* (1939):

> The defendant is charged with an offence against s.58 of the Offences
> Against the Person Act, 1861. That section is a re-enactment of earlier
> statutes, the first of which was passed at the beginning of the last century
> . . . But long before then, before even Parliament came into existence the
> killing of an unborn child was by the common law of England a grave
> crime . . . The protection which the common law afforded to human life
> extended to the unborn child in the womb of its mother. [10]

The traditional law of abortion has, however, undergone a transformation in the last few decades not only because it was ineffective but also
in recognition of the special status and rights of the woman bearing the
fetus. [11] *R v. Bourne* (1939), [12] the English *Abortion Act 1967*, the Menhennit
ruling in the Victorian case of *R v. Davidson* (1969) [13] which was later followed
in the New South Wales District Court's decision of *R v. Wald* (1971), [14]
and the United States Supreme Court decision in *Roe v. Wade* (1973) [15] have
all been instrumental in their respective jurisdictions in effecting change to
make abortion widely available. Despite this seemingly satisfactory position,
abortion continues to be a hotly debated issue, and it has been observed
that, at least in Victoria, abortion law rests shakily on a traditional statutory
foundation which could upturn and upset the balance achieved to date. This
also applies in the United States where there have been recurrent fears that

the Supreme Court will overrule *Roe v. Wade*. To avoid this possibility it has been suggested that a new statute should be enacted incorporating the position represented by the case law.[16]

The description of the offence of child destruction also offers an insight into the working of the criminal law in this area. Section 10 of the *Crimes Act 1958* (Victoria)[17] provides:

> 10. (1) *Any person who, with intent to destroy the life of a child* capable of being born alive, *by any wilful act unlawfully causes such child to die before it has* an existence independent of its mother *shall be guilty of the indictable offence of child destruction . . . [Emphasis added.]*
>
> (2) *For the purposes of this section evidence that a woman had at any material time been pregnant for a period of twenty-eight weeks or more shall be* prima facie *proof that she was at that time pregnant of a child capable of being born alive.*

Two pertinent points should be noted from this quote:

1 the offence of child destruction does not protect all the children *in utero* but only those 'capable of being born alive',[18] and

2 there is a presumption that an unborn child of 28 weeks or more is capable of being born alive.

While the section does not seem to exclude 'children' aged less than 28 weeks from protection, if proof is forthcoming that such 'children' are capable of being born, one could argue that the requirement of such proof and the presumption in favour of those children aged over 28 weeks will in practice protect the latter more favourably. It may further be argued that with modern methods of neonatal care, whereby it is becoming possible to keep alive low birth-weight children, this presumption needs to come down. The requirement of capacity to be born alive has ancient origins in Roman law and was premised on the belief that an unborn child was part of its mother and could not acquire independent rights unless it obtained an independent existence. The English jurist Sir Edward Coke (1552–1634) stated it as follows:

> If a woman be quick with childe, and by a potion or otherwise killeth it in her wombe, or if a man beat her, whereby she is delivered of a dead childe, this is a great misprision . . ., and no murder; but if the childe be born alive and dyeth of the potion, battery, or other cause, this is murder; for in law it is accounted a reasonable creature, in rerum natura, when it is born alive.[19]

The period of 28 weeks seems to have been chosen to mark the time when the child is viable and likely to be born, and thus fill the legal loophole that existed because the offences of abortion and murder did not cover the killing of a child while it was being born.[20]

On the other hand, lowering the 28-week period would upset the purpose of the offence of child destruction, namely, to cover cases where the offences of abortion and murder did not apply, and would undermine the social balance that underpins current abortion law and practice. This stratagem has been made most evident in England where attempts have been made to circumvent the *Abortion Act 1967*. The Infant Life (Preservation) Bill which was introduced in the House of Lords in late 1986 was aimed at reducing the period of 28 weeks given in s. 1 of the *Infant Life (Preservation) Act 1929* to 24 weeks, thus stopping any abortions beyond this period.[21] Another attempt was made in 1987 in the case of C *v.* S[22] where an Oxford student sought to stop his 18-week pregnant girlfriend from having an abortion under the *Abortion Act 1967* by arguing that the fetus was 'capable of being born alive' and that the abortion would contravene the 1929 Act. Both the High Court and the Court of Appeal refused the argument and upheld the woman's choice which was lawful under the 1967 Act.

Outside the criminal law, the determination of the legal status of the embryo/fetus is not so clear either. There have been statutory and court involvement in Anglo-American law in the areas of property law, especially succession, and torts. In the latter area new forms of claims have emerged in the form of, for example, wrongful life, birth and even conception, all of which relate to wrongs allegedly committed on prenates.[23] In England and Australia prenatal rights relating to recovery for injury *en ventre sa mere* (in the womb of its mother) are governed by the 'born alive' rule.[24] This was restated in England by the Law Commission in its *Report on Injuries to Unborn Children* that, in order that there be any cause of action for prenatal injuries, the plaintiff should be a living person and not a fetus, for in the latter state the 'plaintiff has no legal existence at the time of his injury nor has he, prior to live birth, an existence separate from his mother'.[25] The resultant *Congenital Disabilities (Civil Liability) Act 1976* endorsed the 'born alive' rule as have the courts whenever the rights of the unborn were in issue.[26] For example, in *Paton v. BPAS Trustees* (1979) the plaintiff sought to stop his wife from obtaining an abortion without his consent. He argued that he had a right to a say in the destiny of the child he had fathered. In discussing the rights of the plaintiff the question of the rights of the fetus were also discussed. The judge said:

> The foetus cannot, in English Law, in my view, have a right of its own at least until it is born and has a separate existence from its mother. That permeates the whole of the civil law of this country . . . and is, indeed, the basis of the decisions in those countries where law is founded on the common law, that is to say, in America, Canada, Australia . . .[27]

In Australia the 'born alive' rule has also been utilized in at least two different situations. In *Watt v. Rama* (1972)[28] the plaintiff, a three-year-old, brought action for damages for injuries suffered, while *en ventre sa mere*, in a vehicle collision between her mother and the defendant. The defendant argued that at the time of the collision the plaintiff was non-existent as a person, being at that time part of her mother and, therefore, that no duty of care was owed her. The Supreme Court of Victoria found that a duty of care was owed to the plaintiff as a person when born and that it was only as a living person that the plaintiff could sustain injury. In *Kosky v. The Trustees of the Sisters of Charity* (1982)[29] the principle in *Watt v. Rama* was extended to negligent conduct affecting the plaintiff's mother long before the plaintiff's conception and birth. The plaintiff's mother had as a result of negligent conduct on the part of the defendant's hospital been given a wrong blood transfusion which affected the plaintiff when conceived eight years later. The defendant argued that as, at the time of the blood transfusion, the plaintiff was not in existence, no duty of care could be owed him. Although the court was only concerned with the issue of limitation of actions and did not try the case, it found that *Watt v. Rama* was squarely applicable.

Notwithstanding the above authorities, the 'born alive' rule has, in Australia, been interpreted in some cases to mean that a fetus has no rights at all. In *K v. Minister for Youth and Community Services* (1982),[30] the issue arose before the New South Wales Supreme Court as to whether an unborn child could through a representative be party to court proceedings. Street CJ answered negatively, stating: 'I am not, as at present advised, satisfied that the unborn child or foetus has the requisite status to participate as a party in proceedings . . .'. The most explicit rejection of the legal status of the fetus was, however, made in *Attorney-General for the State of Queensland v. T* (1983)[31] where the applicant sought in the High Court of Australia to restrain the respondent, who was allegedly pregnant with his child, from having an abortion. One of the arguments advanced by him was that an unborn child is to be regarded as a person whose existence the courts should protect. In rejecting the application and the argument, Gibbs CJ applied *Paton*'s case and the orthodox rule that a fetus has no rights of its own until it is born and has a separate existence from its mother.

Although the above decisions have been interpreted as deciding that an embryo/fetus is not a person,[32] it must be pointed out that by virtue of reliance on the 'born alive' rule they do not constitute any clear authority on the matter. The rule as it stands has a lot of jurisprudential problems in so far as it seems to equate capacity to exercise rights with attribution of those rights. This problem clearly emerges in those cases where the nature of prenatal rights has been discussed *in extensio* as opposed to those which merely invoked the 'born alive' rule. In the Canadian case of *Dehler v. Ottawa Civic Hospital* (1980),[33] for instance, the foregoing problem was foremost in the judgment of Robins J who, when discussing the rights of the 'unborn child', made the statements in the following two passages:

> *While there can be no doubt that the law has long recognised foetal life and has accorded the foetus various rights, those rights have always been held contingent upon a legal personality being acquired by the foetus upon*

its subsequent birth alive and, until then, a foetus is not recognised as included within the legal concept of 'persons'. [Emphasis added.][34]

Later on the judge stated:

> *In none of the decisions . . . has the foetus been regarded as a person before its birth. In short, the law has set birth as the line of demarcation at which personhood is realized, at which full and independent legal rights attach, and until a child en ventre sa mere sees the light of the day it does not have the rights of those already born.*[35]

The question which is not conclusively answered relates to the nature of the so-called contingent rights. For instance, does the fetus possess rights like a minor or mentally disabled person and merely lack capacity to exercise them? While this appears implicit in the 'born alive' rule, the analogy fails, because a fetus, unlike a minor or mentally disabled person, cannot exercise its rights through a legal representative or its next of kin. If the fetus has no rights at the time when it is a fetus but gets them retrospectively upon live birth, although it is still legally incapacitated in several aspects, then the 'born alive' rule is a jurisprudential curiosity, which appears to be based on a highly controversial definition of 'person' based on Pollock's statement that a 'person is such not because he is human, but because rights and duties are ascribed to him'.[36] Although Pollock could, for instance, only furnish the example of slaves as non-person human beings,[37] his approach could be applied to the unborn, as the 'born alive' rule demonstrates. In addition, the case for the 'born alive' rule is, like in criminal law, weakened by its false premise that a fetus is merely part of its mother. Biologically the fetus is separated from its mother by the placental membranes and is clearly a distinct entity.

In assessing the effect of the above discussion, the position of the unborn in Australia (as well as England) can be stated as follows:

1 In criminal law the unborn child's rights are recognized to the extent that it is protected from destruction (as, for example, sections 10 and 65 *Crimes Act 1958* (Vic.) and corresponding provisions in other Australian jurisdictions). However, abortion law merely affords theoretical protection, while with child destruction the fetus has to be 'capable of being born alive' which is legally presumed as at least 28 weeks. Since the days when the 28-week standard was laid down the technology relating to intensive neonatal care has improved and it continues to improve, so it has been argued that this has rendered the standard inadequate and outmoded.[38]

2 In civil law, especially in torts, the courts have gone a long way towards recognizing fetal life. Although the 'born alive' rule, which is a more strict standard than the criminal 'capable of being born alive' one, is still applicable, especially in the *en ventre sa mere* cases, fetuses have been recognized as possessing rights. In the cases examined, as for instance, *Paton's*

case, *K v. Minister for Youth and Community Services* and *Attorney-General for the State of Queensland v. T* where the full rigour of the 'born alive' rule applied, it could be argued that, because they involved the sensitive question of abortion and the interests of the woman concerned, the precedent value of these authorities should be restricted as regards the status of the IVF fetus. Even with cases involving prenatal injury an argument can be made that the fetus has a special legal place. This could be extended by the advocates of the 'personhood' status of the embryo to mean that, in the face of new IVF technology, the law has an adequate base to protect the IVF embryo against experimentation.[39]

The position of prenatal life in the United States

That the area of fetal rights is in a state of flux is, perhaps, nowhere better illustrated than in the United States. Although this examination can only be cursory and limited mainly to civil law aspects,[40] it will still serve to demonstrate the changes in the law in an attempt to cope with technological, scientific and attitudinal changes, a direction in which Anglo-Australian law may be heading.

In the 1884 case of *Dietrich v. Inhabitants of Northampton*[41] the law relating to recovery for prenatal injury was laid down by Holmes J. The case related to the negligent death of a 'child' who was four months old when its mother miscarried as a result of the defendants' negligence. The court held that there could not be any recovery as at that stage the child could not be considered a 'person' since it did not have an existence separate from its mother. This rule remained virtually unchallenged until the 1940s when courts started to allow recovery for prenatal injuries and introduced the standard of viability.[42] Some courts then found rights of action for wrongful death of a child who was injured *in utero* but was born alive,[43] extended it to an action for personal injuries to a viable child *in utero*,[44] and eventually recognized a right of action for wrongful death of a viable child even though it was not born alive.[45] Thus in *White v. Yup* (1969)[46] (a decision which was cited but not followed in *Paton's* case[47]) the plaintiff sued for $10000 in damages for the wrongful death of her eight-month-old fetus in a car collision with the defendant. The court found the law recognized that a fetus was a 'person' within the *Wrongful Death Act* of Nevada and that an action could be maintained in respect of its death if it were viable at the time. In reaching its finding, the court in an interesting review of the applicable law discussed arguments against allowing recovery. On the 'born alive' argument and its premise that an unborn child is part of its mother, the court stated: 'This proposition has no scientific or medical basis in fact and has been expressly rejected by numerous authorities'.[48]

Viability

I should mention at this stage that the American standard of viability is not free from the problems that have plagued the 'born-alive' rule. These problems, especially the definitional, are highlighted below.

Viability has not been easy to define, especially when its adjective has been used in relation to the terms 'embryo', 'fetus' and 'child'. While 'embryo' has in most, but not all, cases been distinguished from 'viable fetus', 'fetus' seems to have been used interchangeably with 'child'. In *Bonbrest v. Kotz* (1946) (see note 42 above) an embryo was defined as 'the fetus in its earliest stages of development, especially before the end of the third month', while 'viable' was taken as meaning that the 'fetus has reached such a state of development that it can live outside the uterus'.[49] The period when 'such a state of development' would be reached was not indicated. Similarly, in *Wendt v. Lillo* (1960),[50] 'embryo' was defined the same way without any indication of the cut-off point between it and the viable fetus.[51]

Viability has been central in the criminal law as well, especially in the law of abortion. *Roe v. Wade*[52] is founded on it in order to demarcate the borderline between a woman's right to privacy and the so-called State's interest in 'potential life'. In *Roe v. Wade* particularly, a fetus was said to be viable if it was 'potentially able to live outside the mother's womb, albeit with artificial aid'.[53] No fixed period was in principle laid down[54] which, as interpreted in *Planned Parenthood of Central Missouri v. Danforth* (1976),[55] meant that viability was a matter to be determined by medical judgement. In *Colautti v. Franklin* (1979) the principles on viability were restated:

> *Viability is reached when, in the judgment of the attending physician on the particular facts before him, there is a reasonable likelihood of the fetus' sustained survival outside the womb, with or without artificial support. Because this point may differ with each pregnancy, neither the legislature nor the Courts may proclaim of the elements entering into the ascertainment of viability, be it weeks of gestation or fetal weight or any other single factor . . .*[56]

Although the law strives to be definite and certain, a rigid application of viability, especially outside the abortion cases, could stifle its development. If viability was fixed at eight or even seven months (as it was in some cases) it could resemble the 'capable of being born alive' standard and lead to similar consequences. In its present flexible form the test of viability could actually support the case of those who think that the embryo should be accorded legal protection.[57]

Fetal experimentation

The issue of experimentation, whether on an IVF embryo or embryo/fetus *in utero*, appears to have emerged as part of the general question of the rights of the unborn as well as the issue of experimentation on human subjects. While I do not intend here to delve into the law of human experimentation, it must be noted that, although the controversy surrounding fetuses including embryos is distinctly separate from human experimentation, the definitional approach towards both appears to be the same.[58] It needs to be stated at the outset that the law relating to experimentation on the fetus or even human beings is very scanty. The cases of *Slater v. Baker and Stapleton* (1767)[59] and *Halushka v. University of Saskatchewan* (1965)[60] which are often cited as experimentation cases[61] establish experimentation as a form of negligent conduct and do not define it adequately outside the context of informed consent. More significantly the decisions are not concerned with the fetus but with mature human beings.

In the absence of clear legal authority, one view could be that debate on experimentation is open to philosophical, scientific and, sometimes, religious arguments on the moral status of the fetus/embryo. In this debate experimentation is either justified or not depending on whether the view-point taken accords a moral status to the embryo/fetus.

Another view could flow from an extension of the principles discussed above—that if the law is interpreted as moving towards a greater recognition of the rights of the fetus, this recognition should be extended to all forms of the fetus, including the embryo. As the development of research on the fetus is of recent origin, the law should broaden existing principles to deal with an aspect created by advances in reproductive technology. As far as civil liability is concerned, if an embryo is damaged during experimentation and a resultant child suffers any defects from such damage, existing principles clearly favour recovery against the person(s) experimenting.[62] Even with the rigid 'born alive' rule, liability may still attach for destroying an embryo despite the problem of proof of damage which may be involved. In the only known IVF case of *Del Zio and Del Zio v. The Presbyterian Hospital* (1978)[63] the plaintiffs whose embryo was destroyed by the defendants were held to be entitled to damages for emotional distress. The defendants had argued unsuccessfully that the *in vitro* experiment was yet to be perfected and that in its present state could lead to malformation and damage to any resultant life. The precedent value of the *Del Zio* case is not immediately clear, as the court did not at all discuss the issue of the status of an IVF human embryo (including whether its destruction *per se* was wrong) nor, indeed, whether the embryo was the property of the Del Zios (the claim against conversion was rejected by the jury as it found the embryo's monetary value could only be speculative). However, the case could still be some authority for the protection rather than the destruction of fetal life.[64]

In the United States fetal research has been considered at length since the 1970s[65] and some states have enacted legislation prohibiting it whether in relation to fetuses *in utero* or *in vitro*.[66] Although the overall legal and policy position is rather confused, there appears to exist a moratorium on IVF federal funding which has in turn affected experimentation.[67]

Conclusions

As I said at the beginning of this chapter, the legal world has greeted the advent of IVF human embryo reproductive technology with anguish. Not only are 'capable of being born alive' and 'born alive' concepts rather out-dated, but also the law has sometimes been evasive and rather apprehensive when issues involving the status of prenatal life have gone before the courts.

While in several instances prenatal life seems to have been recognized in the offences of child destruction, abortion and, in some cases, murder[68] and recovery of damages for prenatal injury and wrongful death, a compelling conclusion is that this recognition has no firm principle behind it. However one extends its significance, it cannot supply the necessary basis for regulating the problems posed by the new IVF technology.

The specific question of IVF human embryo experimentation cannot, therefore, find its answer in the common law and traditional statutory law but must be tackled afresh, as various committees have recommended, with due recognition of the sensitivities and dilemmas involved.

Notes

1 See L. Walters, 'Ethics and new reproductive technologies: an international review of committee statements', *Hastings Center Report* Special Supplement 17(3) (1987), 3.

2 Studies in embryology do not endorse this classification and exclude from the embryo stage the first two weeks of cell division: see e.g. A. McLaren, 'Why study early human development?', *New Scientist*, 24 April 1986, 49.

3 See L. Waller, 'Any reasonable creature in being', *Monash University Law Review*, 13 (1987), 37.

4 (1973) 410 US 113 at 159 by Blackmun J.

5 [1979] 1 QB 276 at 278.

6 (1977) 67 111 2d 348 at 353. See also *Bennet v. Hymers* (1958) 147 A 2d 108.

7 [1972] VR 353 at 377.

8 The problems associated with defining the beginning of a human being are considered in C. Howard, *Criminal Law*, 4th ed. (Law Book Co., Sydney, 1982), pp. 23–5.

9 See ss 10 and 65, respectively of the *Crimes Act 1958* (Vic.). In some other Australian states provisions relate to child murder and infanticide: see e.g. ss 20 and 22A of the *Crimes Act 1900* (NSW).

10 [1939] 1 KB 687 at 690 by Macnaughten J.

11 Some of the factors militating against the traditional law of abortion are discussed in B. Dickens, *Abortion and the Law* (MacGibbon & Kee, London, 1966), pp. 107–71.

12 [1939] 1 KB 687.

13 [1969] VR 667.

14 (1971) 3 DCR (NSW) 25.

15 (1973) 410 US 113.

16 Waller, op. cit., 53–4.

17 See, too, the *Criminal Code 1913* (WA), s 290 and the *Criminal Code 1924* (Tas), s 165.

18 This standard is distinguishable from and not related to the various criteria for determining birth as, for example, breathing or crying, to be found in legislation on the registration of births.

19 3 Coke, *Institutes* 58 (1648) cited in M. Barrazotto, 'Judicial recognition of feticide: usurping the power of the legislature?' *Journal of Family Law* 24 (1985), 43, 45.

20 See e.g. Victorian Parl Debs, 1949, Vol 229 at pp. 902, 1243, 1251 and 1252 during the debate on cl 5 of the Crimes Bill 1949, which later became the present s 10 of the *Crimes Act 1958*. Although G. Wright, 'Capable of being born alive', *New Law Journal* 131 (1981), 108, has argued that viability is not envisaged in s 1 of the English *Infant Life (Preservation) Act 1929*, and the Victorian section is clearly copied therefrom, the debate in the Victorian Parliament interpreted 'capable of being born alive' as including a 'viable child' (ibid. at p. 1243).

21 *International Medical Ethics Bulletin*, 21 (December 1986), 15.

22 [1987] 2 WLR 1108.

23 See, e.g., B. Dickens, 'Wrongful birth and life, wrongful death before birth and wrongful law' in S. McLean (ed.), *Legal Issues in Human Reproduction* (Gower Publishing, Aldershot, England, 1989), p. 80.

24 The rule's rather curious origins reach as far back in history as 1694: see R. Pound, *Jurisprudence*, 4 (West Publishing Co., St Paul, Minn., 1959), pp. 388 et seq.

25 Law Com No 60 (Cmnd 5709), 1974, par. 33.

26 The Act is discussed in, e.g., P. Pace, 'Civil liability for pre-natal injuries', *Modern Law Review* 40 (1977), 141, 149–58.

27 [1979] 1 QB 276 at 279. The law in America, as will be shown below, is different on this aspect.

28 [1972] VR 353.

29 [1982] VR 961.

30 (1982) 8 Fam LR 250.

31 (1983) 57 ALJR 285.

32 See, e.g. Russell Scott, submission to Senate Select Committee on the Human Embryo Experimentation Bill 1985, Senate Hansard Report, 23 April 1986, 1317, p. 1338.

33 (1980) 101 DLR (3d) 686.

34 Ibid. at 695.

35 Ibid.; see also at 699.

36 Robins J in the *Dehler* case, ibid., at 695, relied on such a definition from the 2nd edition of Pollock's *First Book of Jurisprudence* (Macmillan & Co., London, 1904), pp. 110–11.

37 Pollock, ibid., p. 111.

38 The question of resources to maintain these babies, however, continues to be debated: see, e.g. P. Pemberton, 'The tiniest babies—can we afford them?', *Medical Journal of Australia* 146 (2) (1987), 63.

39 During the hearings of the Australian Senate Select Committee on the Human Embryo Experimentation Bill 1985 the above authorities, as for example, *Watt's* and *Kosky's* cases, were cited in support of the Bill: see Right to Life Association (New South Wales Branch), submission No 180, Senate Hansard Report, 26 May 1986.

40 The status of the embryo/fetus under criminal law is not the same throughout the various states. While in some states homicide statutes have been amended or even interpreted by courts to include the prenate, many others still lack these legislative and judicial overtures: see T. Mallory and K. Rich, 'Human reproductive technologies: an appeal for brave new legislation in a brave new world', *Washburn Law Journal* 25 (1986), 459, 466–8.

41 (1884) 52 AM R 242.

42 *Bonbrest v. Kotz* (1946) 65 F Supp 138 is regarded as the first case to have swung the trend away from *Dietrich's* case while *Keyes v. Constructions Serv Inc* (1960) 340 Mass 633 has been interpreted as having effectively overruled *Dietrich's* case: see *White v. Yup* (1969) 458 P 2d 617 at 620.

43 See e.g. *Amann v. Faidy* (1953) 415 Ill 422 and *Woods v. Lancet* (1951) 303 NY 349.

44 See e.g. *Rodriguez v. Patti* (1953) 415 Ill 496.

45 See e.g. *Verkennes v. Corniea* (1949) 229 Minn 365 and *Chrisafogeorgis v. Brandenburg* (1973) 304 NE 2d 88.

46 (1969) 458 P 2d 617.

47 [1979] 1 QB at 279. It was cited as the 'only' departure from the 'born alive' rule known to Sir George Baker P.

48 (1969) 458 P 2d 617 at 623.

49 (1946) 65 F Supp 138 at 140.

50 (1960) 182 F Supp 56.

51 Ibid., at 57. See, too, *Mitchell v. Couch* (1955) 285 SW 2d 901 at 905.

52 (1973) 410 US 113.

53 Ibid., at 160.

54 In *Roe v. Wade* (ibid.) 28 weeks was taken as the viability stage although the court noted that it could even be 24 weeks.

55 (1976) 96 S Ct 2831 at 2838–9.

56 (1979) 99 S Ct 675 at 682.

57 In the area of abortion it has been argued by, e.g., R. Blank, 'Judicial decision-making and biological fact: Roe v. Wade and the unresolved question of fetal viability', *Western Political Quarterly* 37 (1984), 584, that because of the heavy reliance on viability, *Roe v. Wade* may actually become an anti-abortion ruling.

58 This refers to the tendency of classifying experiments as either therapeutic or non-therapeutic.

59 (1767) 95 ER 860.

60 (1965) 53 DLR (2d) 436.

61 See C. Levy, *The Human Body and the Law: Legal and Ethical Considerations in Human Experimentation*, 2nd ed. (Oceana Publications, London and New York, 1983), p. 65, and B. Dickens, 'What is a medical experiment?', *Canadian Medical Association Journal* 113 (1975), 635.

62 See, too, G. Smith II, 'Australia's frozen orphan embryos: a medical, legal and ethical dilemma', *Journal of Family Law* 24 (1985), 27, 35.

63 US District Court, SDNY 1978 74 Civ 3588 (CES) reprinted in *Bioethics Reporter* 2 (1985), 7.

64 Interpretation of the decision by writers has varied between those who see it as conferring 'a legal value if not a legal status of some sort on the embryo' (see R. Blank, *Redefining Human Life: Reproductive Technologies and Social Policy* (Westview Press, Boulder, Colorado, 1984) p. 161) and others who find it inconclusive (see, e.g., J. Saltarelli, 'Genesis retold: legal issues raised by the cryopreservation of preimplantation of human embryos', *Syracuse Law Review* 36 (1985), 1021, 1047–8).

65 See, e.g., D. Horan, 'Fetal experimentation and federal regulation', *Villanova Law Review* 22 (1977), 325.

66 For a list of states which have passed such legislation, see N. Terry, '"Alas! poor Yorick," I knew him *ex utero*: the regulation of embryo and fetal experimentation and disposal in England and the United States', *Vanderbilt Law Review* (1986) 419, 462–6.

67 See, e.g., L. Kass, 'The meaning of life—in the laboratory' reprinted in *Bioethics Reporter* 2 (1985) 221, 223.

68 See, e.g., *The People v. Apodaca* (1978) 142 Cal Rptr 830, where the defendant was convicted of the murder of a fetus that was between 22 and 24 weeks old.

15 | Self-regulation and embryo experimentation in Australia: A critique

PASCAL KASIMBA

The question of what form IVF regulation should take is emerging as a rational consideration from the rather emotional excitement that has characterized the response to embryo experimentation. This chapter analyses self-regulation to see how far it is a satisfactory approach to regulating embryo experimentation. In order to examine the problem I shall define self-regulation, present and appraise the self-regulation structure in Australia and analyse the practical and theoretical arguments that surround it. Although the focus is on Australia, the issues to be examined are universal.

Before proceeding, a few broad observations need to be made. Although the predominant issue is human embryo experimentation, the self-regulation structure that has evolved in Australia has largely been in the area of human experimentation. It is this general scheme that some seek to extend to the embryo. While this may be a good starting point to assess the record of self-regulation, we need to keep in mind the shifting moral bases that underly *human* and human *embryo* experimentation. It is also worth observing that the ethical codes, guidelines and behaviour that form the core of the self-regulation approach revolve around the professional medical practitioner. This relationship should be kept in mind since the issue of embryo experimentation, perhaps more than human experimentation, is outside the doctor–patient relationship and encompasses researchers from other scientific disciplines.

What is self-regulation?

Although one hears of suggestions that embryo research should be left to self-regulation, the meaning of 'self-regulation' frequently varies. So what exactly is self-regulation? How does it apply when different professions (e.g. medical and other scientists) are engaged in human embryo research?

The ordinary meaning of the term 'self-regulation' implies that one controls oneself to the exclusion of anyone else. Applied to activities of groups or organizations, self-regulation would mean that controls are imposed from within the group or organization, not by outside intervention. The use of the term in the medical and scientific contexts does not, however, conform to its ordinary meaning. Sometimes it is equated with 'peer review' to describe the checks that, allegedly, already exist in the conduct of IVF research. This essentially is taken to mean that control is exercised by the same scientific group which may even exist in a licensure framework and can censure one of themselves for any departure from 'accepted' standards. This concept of self-regulation is borrowed from the context of the medical profession where it has been described thus:

> The field of medicine is highly self-regulated. Although medical licensure procedure is authorized by a state legislature, it is implemented and controlled by the profession. Medical licensure therefore offers a well-developed model of how a profession can self-regulate, particularly since those in the field are highly motivated toward self-improvement by means of private certification and since the patients are obvious benefactors of this self-advancement. Hospitals also play a major role in regulating physicians, since doctors must be approved for hospital affiliation and this affiliation is extremely important for doctors to practice medicine profitably. Peer review, therefore, is constantly in effect throughout the career of physicians. [1]

Another meaning of self-regulation refers to the capacity of individual IVF scientists to regulate the conduct of experimentation in accordance with their own ethical perceptions. In the area of human experimentation this conception of self-regulation has been widened to rationalize limitations imposed by institutional ethics committees (IECs). According to one commentator,

> the medical profession, or at least those most caught up in experimentation, will prefer to devise and operate their own internal controls . . . These professional controls are of two kinds, those which the experimenter imposes on himself—the internalized controls of conscience and the acceptance of socially required limitations on his conduct—and institutional committees whose approval must be secured before the experiment goes forward. [2]

The nature of the 'self-regulation' described in the wide sense seems to depend on the composition of the institutional ethics committees. Some proponents of peer review regard it as essential that scientific and medical research should be judged exclusively by scientific and medical committees. They have argued that lay representation on committees, like the institutional review boards (IRBs) which operate in the United States, greatly reduces the advantages of peer review.[3]

In Australia the area of human embryo research has since 1982 been the subject of guidelines issued by the national body that oversees medical research—the National Health and Medical Research Council (NH & MRC). Although the NH & MRC has denied that its regulatory structure should be described as 'self-regulatory'[4] it is reasonable to infer that when it is said that IVF research in Australia is self-regulatory, the NH & MRC's guidelines are given a dominant place in the regulatory mechanism(s).[5] It may even be said that the basis for the NH & MRC's role in this area fits within the vein of self-regulation, as the brief history of its ethical regulation will reveal.

The NH & MRC and the regulation of human embryo research

The NH & MRC's ethical regulation dates from 1966 when, in its capacity as a national body engaged in funding scientific and medical research in Australia, it issued guidelines on human experimentation.[6] Institutional ethics committees were introduced much later in the 1976 NH & MRC Statement on Human Experimentation. This contained the requirement that research proposals for NH & MRC grants should not only accord with certain ethical principles but also be approved by the institutional ethics committee of the institution where human research was to be done. The purpose of the 1976 additions was to strengthen peer review in the face of increasing pressure from many sectors.[7] In 1982, following the *Report on Research on Humans* by the NH & MRC Working Party on Ethics in Medical Research, the guidelines were overhauled to meet criticisms of, for example, lack of community involvement and the 'need to keep research workers sensitive to community attitudes'.[8] Supplementary Note 4, containing specific guidelines on IVF, was introduced at this time which also saw the creation of a central ethical body, the Medical Research Ethics Committee (MREC).[9]

The regulatory structure

The institutional ethics committee concept occupies a central place in the NH & MRC's regulatory structure. As far as human embryo research is concerned, the NH & MRC's Supplementary Note 4 on *In Vitro* Fertilization and Embryo Transfer is of substantial importance. In addition there is the NH & MRC Statement on Human Experimentation which forms the background of this regulatory structure embodied in Supplementary Notes 1 and 4. The relationship of all these notes and the statement is outlined in Supplementary Note 4 which requires that the Statement on Human Embryo Experimentation should apply to work on IVF and that centres with IVF and ET programs should have all aspects of their work approved by an IEC.

The functions of an IEC regarding IVF are set out in Supplementary Note 1 and Paragraph 1 of Supplementary Note 4 (herein called the IEC note and IVF note respectively). Briefly, the IEC note lays down the composition and functions of IECs. Two of the functions are 'to consider ethical implications of all proposed research projects and to determine whether or not they are acceptable on ethical grounds' and to 'keep the progress of research projects under surveillance so as to be satisfied that they continue to conform with approved ethical standards' (para. 4(i) and (ii)).

It is not clear from the guidelines in both the IEC and IVF notes how the IEC is to perform its duty of vetting IVF research. The NH & MRC Statement on Human Experimentation (which has to be complied with by virtue of the IVF note) provides that research on human beings should conform to 'generally accepted moral and scientific principles' (para. 1). Apart from the fact that the Human Experimentation Statement was designed to apply to human beings, the statement does not indicate what these 'moral principles' may be, except for the preambular reference to the Declaration of Helsinki, 1964, as revised, and the Proposed International Guidelines for Biomedical Research Involving Human Subjects, 1982, which are 'intended as a guide on ethical matters bearing on human experimentation'. If, however, the ethical principles contained in these instruments are included in the 'moral principles' of the Human Experimentation Statement, they conflict with Para 5 of the IVF note, which provides:

> *Research with sperm, ova or fertilized ova has been and remains inseparable from the development of safe and effective IVF and ET; as part of this research other important scientific information concerning human reproductive biology may emerge . . .*

So what is the IVF note intended to mean when in its opening paragraph it states that 'while IVF and ET is an established procedure, much research remains to be done and the NH & MRC Statement on Human Experimentation should continue to apply to all work in this field'? In other words, how can the IVF note juxtapose the application of ethical principles about human experimentation to embryo research with a value position that denies categorically that a human embryo is a 'human'? While human experimentation principles place emphasis on the interests of the subject, embryo research includes destructive research which clearly goes against the embryo's interests.

In a submission to an Australian Senate Select Committee, the NH & MRC's Medical Research Ethics Committee restated the position that underlines the IVF note:

> *We believe that the relevant ethical question here is not 'When does life begin?' but 'What degree of protection is due to the fertilized ovum, the embryo, and the fetus at the succeeding stages of its development?' The MREC's conclusion, after considering all the views conveyed to it, is that, up to the stage when implantation could occur, the fertilized ovum should not necessarily be regarded as having a status which calls for such protection as would prevent it being the subject of experimentation.* [10]

How have IECs actually done their work? Although the IECs in Australia were given the IVF task in 1982 there is hardly any literature on how they have handled it, particularly the protocols that relate to embryo research.[11] While the IEC note lists, as one of the functions of an IEC, keeping a record of all proposed research projects, such a record does not have to, and probably does not, contain reasons for ethical decisions.[12] Further, Paragraph 1 of the IVF note extends the status of 'medical record' to such IEC records so that they are removed from the public scrutiny by the doctrine of confidentiality.

The MREC has, however, produced a report, *In Vitro Fertilization Centres in Australia: Their Observance of the National Health and Medical Research Council's Guidelines* (August 1986). The MREC set out to audit IVF centres (thought to be twelve in number) and assess their observance of the IEC and IVF notes. The MREC met members of IVF teams and IECs in those institutions with IVF programs. The report covers two main aspects—compliance with the IEC and the IVF notes. Briefly the main findings were that three of the twelve centres visited did not follow the IEC note regarding composition of the committee. Two of these had too few lay representatives.[13] Further, the report found that although IECs complied with the record-keeping requirement, some of these records were not sufficiently detailed. Hence the report recommended that 'institutional ethics committees should record the basis for decisions made'.[14]

The report notes that generally the guidelines in the IVF note were observed by IVF centres. For instance, they kept records, accepted only married couples and did not approve any projects that would lead to the growth of an embryo beyond the implantation stage. However, two main problems, one relating to the interpretation of 'research' and 'experiment' and the other to the adequacy of consent forms, were encountered. The inadequacy of the guidelines on human embryo research was manifested in the varied way that IVF centres interpreted it. In identifying this problem the MREC reported:

> *We took a wide view of the scope and meaning of the expressions 'embryo', 'research' and 'experiment'. Medical and scientific*

understanding of the definition of 'embryo' is not unanimous, and opinions differ about what amounts to 'research' and 'experiment'. We therefore inquired beyond the recorded research projects at each institution. It became apparent that some centres held the view that certain activities did not amount to research, or at least to the kind of research that requires the specific approval of an institutional ethics committee. A common activity is one which involves changing the ingredients which comprise the culture medium or nutrient solution in which fertilized ova are placed prior to transfer. Another is variation in the method of manipulating the fertilized ovum. Most centres held the view that minor changes and variations in these procedures did not amount to research or experimentation or, at least, did not require specific institutional ethics committee approval. We consider, however, that all such changes and variations do amount to research because they aim to permit valid conclusions to be reached on whether what is being tried is beneficial. The hypothesis is that the new measure will benefit, or at least, not harm, the fertilized ovum. However, the result will not be known until the experiment is done. In order to be ethically acceptable, such activities must be scientifically valid and as well comply with other ethical criteria. [15]

Although the MREC went on to recommend that culture medium or nutrient solution modification should be regarded as research on an embryo and accordingly be 'brought to the notice of the institutional ethics committees', it is not clear what the IECs are supposed to do. If this means that they are to consider them as proposals to be approved as 'ethically acceptable', the problem regarding the ethical criteria that are to be applied still remains. The MREC states, in regard to the activities that amount to research, that 'the hypothesis is that the new measure will benefit, or at least, not harm, the fertilized ovum'. Is this a benefit/harm criterion that should apply to all research on the human embryo? If it is, it would clearly be in conflict with Paragraph 5 of the IVF note and the general position of the MREC and NH & MRC that an embryo should not be exempt from destructive experiments.

The second major problem encountered by the MREC in their audit of IVF centres relates to consent forms and the wishes of donors. The report notes that 'consent forms did not always list the possible options for the donors—that the embryos should be implanted in the donor's womb, donated to another woman for implantation in her womb, used for research or discarded'. [16] Some of these problems spring from the IVF note itself which does not give any model consent form and speaks generally of ascertaining 'the wishes of the donors regarding the use, storage and ultimate disposal of the sperm, ova and resultant embryos' (para. 6).

Apart from the problems unmasked by the MREC audit, the overall regulatory mechanism represented by the NH & MRC guidelines poses yet other problems that revolve around the audit system and the representativeness of the NH & MRC guidelines.

The first group of problems relates to the status and efficacy of the so-called audit. As mentioned above, IECs and IVF centres operate in an atmosphere of secrecy. In answer to the charge that it may not be possible to know what IVF centres and IECs do or how they do it, the MREC has stated that as part of its terms of reference it conducts audits on such centres as well as keeping an eye on IECs. But how is the MREC to overcome the problem or barrier of confidentiality? Confidentiality is a legal obligation so that IVF scientists in a doctor–patient relationship are under a legal duty not to disclose information about their patients to others.[17] While the MREC in its audit of 1985–6 did not appear to encounter such problems, it may be suggested that disclosure of such personal information (for example, details of ovum recovery, fertilization, pregnancy and consents to IVF procedures) may be illegal, or that a possibility exists that the MREC may be refused access to such information on grounds of confidentiality. It has also been suggested that the validity of the audit cannot be vouched for since the information available to the MREC was volunteered. Furthermore IECs have no way of ensuring that the NH & MRC guidelines are followed and that they are not given inaccurate information by IVF researchers.[18]

Another problem with the MREC relates to its status. It has been said that the MREC and IECs are not representative of the wider community and that, in particular, the NH & MRC guidelines, including the IVF note, did not result from any community consultation and were not drawn up by individuals with any semblance of community representativeness. This problem is, perhaps, a serious one if the premise taken is that IVF research is a community rather than a medical or scientific issue. To the extent that the IEC note has evolved away from a strictly peer committee to one with lay representatives, the NH & MRC guidelines may be said to have accepted, at least in principle, that there is a genuine role in the IVF area for community representation. However, the MREC cannot be seen in a similar light to the IEC on the question of representation, although the IEC is itself far from representative. If the MREC wants to express the democratic principle it may have to be replaced by a more representative body. Otherwise, however workable and representative of lay interests the NH & MRC guidelines may seem, it cannot overcome or answer the charge of lack of community representation. The members of the NH & MRC, MREC and IECs are appointed by invisible authority and are not elected representatives.

Apart from the NH & MRC, the Fertility Society of Australia (described as composed of 'the professionals in the business of the fertility units, whether they be scientists, doctors or nurses'[19]) has also issued *Programme Standards for Infertility Units Using in Vitro Fertilization (IVF) and Related Technologies Involving Egg and Embryo Collection and Transfer Including Gamete Intra Fallopian Transfer (GIFT).*[20] These standards require, *inter alia*, that IVF be practised within the framework of the NH & MRC guidelines, that only properly qualified personnel be engaged in it, and that there be counselling, audit by the Fertility Society and submission of IVF pregnancy results to the National Perinatal Statistics Unit at the University of Sydney.

This society and its standards reflect the response of the IVF research community to the IVF debate and show that it is in control of the new IVF technology. However, it also suffers from the problems of the NH & MRC guidelines, not least because it adopts them. Worse than the NH & MRC, however, it is a voluntary association that exhibits the characteristics of peer review when the general trend is to involve lay interests. The society is by and for IVF practitioners and, while this does not necessarily impugn the very high standing in the community that these individuals enjoy, it puts them in the position of being judges in their own cause. It may be desirable for such an association to exist, but if problems of elitism and appropriation of an issue of community concern are to be avoided, such bodies cannot be offered as exclusive alternatives for regulating IVF embryo research.

To summarize this section, the self-regulation structure in Australia has the NH & MRC at its apex. Within this structure come the IECs. Aside from the NH & MRC but on a lower plane sits the Fertility Society of Australia which has endorsed the NH & MRC guidelines for its members.

General overview of the self-regulation concept

Objections that have been raised against self-regulation in general relate to the adequacy of the NH & MRC guidelines in regulating IVF embryo research, lack of representativeness, issues of commercialization, enforcement, cost and the institutionalization of IECs.

The NH & MRC regulatory structure, as was briefly mentioned above, has been accused of, among other things, lacking proper ethical guidelines to regulate research. It is argued that although the NH & MRC guidelines require researchers to submit IVF research protocols to IECs, there is no indication of what ethical guidelines the IECs should follow other than the NH & MRC's own position that pre-implantation embryos can be subjects of research, including destructive research. One commentator has remarked in this respect that 'the apparent inability of the NH & MRC to address the ethical issues involved must cast some doubt on its ability to "regulate" experimentation of this kind'.[21] As we have seen, this criticism is quite serious since the issue is human embryo research and the NH & MRC guidelines adopt a definite position. This could create a practical problem for some IVF researchers who may feel that observing the guidelines may prejudice their chances of getting a more community-based recognition for their work. During the hearings of the Senate Select Committee it was revealed that in May 1986 the IEC at the Royal North Shore Hospital in

Sydney authorized the fertilization of previously frozen oocytes for examination, with no intention of implanting them. The IVF group at the hospital refused to do these experiments because they found it ethically unacceptable to collect eggs, fertilize them and destroy them when there were couples in need of them.[22]

It has also been stated that because the ethical values involved in IVF human embryo research change from time to time and from individual to individual, IECs do not present an adequate guide especially to an IVF group that deals with several institutions with separate IECs.[23] This problem is also canvassed in the context of the federal-versus-state approach. The local IEC is not, however, without support. One writer has suggested that such IECs have

> *greater familiarity with the actual conditions surrounding the conduct of research . . . can work closely with investigators to assure that the rights and welfare of human subjects are protected and, at the same time, that the application of policies is fair to the investigators.*[24]

Self-regulation as it is put forward by the NH & MRC has also been criticized for being elitist, secretive and lacking in representation of women. One commentator has criticized the NH & MRC guidelines and made the accusation that 'the history of self-regulation in medicine has been one of secretiveness based on a superiority complex and resentment when the community has sought to penetrate the veil'.[25] Another has stated that self-regulation cannot be adequate 'given the high concentration of experts and men in the various ethics committees'.[26] These charges are debatable but what may be established is that the community's attitude to workings of professional groups like doctors is changing. In particular, awareness of biomedical research issues seems to be increasing. In the context of IVF, this is evidenced by the concern raised by certain interested groups like women, infertile people and politicians. The Combined New South Wales IVF Support Group, in its submission to the Senate Select Committee, stated that:

> *self-regulation does not allow for representation of patients, their potential children or the community—self-regulation does not allay community fears about possible excesses or aberrations which may arise from unfettered research.*[27]

The *bona fides* of IVF scientists in regulating themselves has also been questioned in view of the presence of vested interests. With the formation of IVF (Australia) Ltd, and other commercially oriented companies now marketing IVF technology in the United States and other countries, it has been said that the secrecy that normally surrounds scientific research has been reinforced by commercial secrecy. It is further argued that the IVF scientists now have a motive not to reveal what they are doing and, even, to do anything in pursuit of career and profits. The adequacy of the NH &

MRC guidelines in this commercial environment is doubted on the basis of lack of enforcement. It is also pointed out that the power of the NH & MRC is related to its funds, and scientists whose grants are from other sources are not regulated by the NH & MRC. The IECs have no way of knowing what research these scientists are doing.[28] Some of these criticisms are admitted by the NH & MRC. In the First Report by the NH & MRC Working Party on Ethics it was stated:

> *While the NH & MRC is an accepted authority in ethical matters bearing on medical research in Australia, it might be said that it is a 'paper tiger', for it has no power to apply the principles and enforce the procedures that it advocates, except in relation to research with which it is itself directly concerned by virtue of funding . . . In this connection we have noted the estimate that the Medical Research Endowment Fund, which the NH & MRC administers, provides little more than one quarter of the total funds spent per year on medical research and development in Australia.[29]*

The NH & MRC, however, hopes to secure wider observance for its guidelines through cooperative arrangements with other major granting bodies. The Heart Foundation and cancer societies, for instance, include in their grant application forms the condition that the proposed research must conform to NH & MRC guidelines.[30] The NH & MRC also has strong links with state Health Departments and the standing committee of Attorneys-General, which would ensure observance at the state level. Similar arrangements with the private sector are also possible. These arrangements, however, depend on what has been termed an 'honour' system, and do not allay the fears about the 'enthusiastic scientist' who does not need funding.[31]

Although these criticisms may not be fair to the average IVF scientist/practitioner, the problem of impotency in enforcing guidelines is a practical and real one which can be illustrated by the experience of the Voluntary Licensing Authority (VLA) set up in Britain by the Medical Research Council and the Royal College of Obstetricians and Gynaecologists in the aftermath of the Warnock Report. According to some reports,[32] the VLA clashed with an infertility clinic at the Wellington Humana Hospital in London over a new VLA guideline that forbids the use of gametes donated by close relatives. The hospital has an ethics committee but the clinic has opposed the guideline and argued that doctors should decide for their patients and that the guideline should be redrafted accordingly. In Australia, IVF clinics in Western Australia disregarded decisions of IECs against transferring multiple embryos that have led to multiple IVF births.[33] The West Australian Government is now intending to introduce legislation to limit the numbers of embryos used. Although these examples could be seen as isolated cases, they might lend some support to the argument that guidelines cannot ensure the necessary compliance and that resort should ultimately be had to legislation.

Further considerations of the self-regulation structure arise from the experiences in other countries. A problem that has not been given much attention locally is that of the cost of an IEC system. In the United States the issue has been considered in relation to institutional review boards, and it has been observed that hospitals and other institutions carrying out human research have limited funding which affects the work of IRBs that process a lot of research proposals.[34] As IECs multiply here and are given new tasks, for instance IVF research protocols, the cost factor will have to be considered as it has a direct bearing on the efficacy of IECs. It may, for instance, have to be considered whether to remunerate IEC members, to employ research officers and provide adequate facilities to allow an informed approach to the consideration of research proposals.

Another problem relates to the clarity of guidelines which the MREC has found lacking on the fundamental issue of what research needs or does not need IEC approval. The problem of vagueness, which has also plagued the guidelines for IRBs in the United States, has focused the necessity for safeguards for sound decision-making. What can a researcher do if he/she feels unhappy with such a decision? Should challenges be mounted in courts? In the United States it has been suggested that another body should review the decisions of IRBs.[35] This in turn has sparked counter-arguments, such as that this will bureaucratize an otherwise non-contentious area and undermine the authority of IRBs if a previously passed protocol is reversed by the review body. This, it is argued, will affect the morale of IRB members as well as lowering the esteem of the system in the eyes of researchers.[36] It is also contended that review is based on a misconception of the role of IRBs, which is not to determine the merits of a research proposal in accordance with some ascertainable rules but to review its ethical standing, something that does not admit of objective assessment—the IRB is not a 'deputy sheriff' which has to regulate research.[37] If we apply these observations to the IEC in Australia, is there any role for a body to review IEC decisions?

Although the NH & MRC system has been the major focus of analysis, it should not be forgotten that it builds on the professional controls that IVF researchers exercise in the day-to-day practice of their work. But who are these researchers and what clues are there to what they apply as controls against, possibly, morally outrageous experimentation? It is well known that the field of medicine contains some ethical prescriptions dating many centuries back to the Hippocratic Oath. Although these are, arguably, applicable to human embryo research, inconsistencies exist, as we have seen in connection with the NH & MRC guidelines. The problem, however, remains when disagreement is encountered on the disciplines involved in IVF. One IVF medical scientist has written that:

> one hopes that enough politicians will remember that infertile couples constitute an appreciable fraction of young voters to reverse the present trend towards *regulating the* delicate and private matter of making medical decisions.[38] *[Emphasis added.]*

Implicit in this quote is the view that IVF is a medical matter which doctors are able, within their professional conduct, to handle well. According to another commentator,

> there are probably several reasons that [sic] self-regulation appears to be working effectively as a risk reducer in the medical field. Usually the patient–doctor relationship is clear, and it is easy to identify irregularities or departures from an expected standard of care.[39]

Such views, however, seem fallacious or inapplicable to IVF. IVF is not exclusively a medical matter and there is disagreement as to whether infertility is even a medical condition. Two IVF practitioners identified the departure from medicine when they wrote:

> In-vitro *fertilization has brought to gynaecological practice a multidisciplinary aspect which includes specialist infertility nursing, social work or social psychology, patient self-help support groups and basic science . . . This is a dramatic departure from the usual practice of medicine where normally a single specialist determines treatment in consultation with the patient and other specialists.*[40]

What can be gathered from this is that IVF is a broad issue that transcends the medical field. As such its regulation cannot be left to the medical profession alone but has to take into account the other disciplines involved as well as rising community concern. Since IVF has become public, it must be demonstrated how this 'self-regulation' will be achieved.

Other problems regarding self-regulation centred on the IEC that have been encountered overseas include technicality of proposals that render them too hard to be assessed adequately and monitoring to make sure that research continues in accordance with approved standards.[41] It has also been pointed out that the IEC system has led to some evils, including the abuse of IEC approval which has obtained an 'intrinsic value' of its own to be used to obtain funds and publications. It is also feared that the system leads to 'institutionalization of ethics' which leads to shared responsibility that falls nowhere. The scientist in a way transfers ethical responsibility to the IEC and the IEC trusts that the researcher will be responsible, comparable to a situation of superior orders where the person giving orders does not act while the actor does not give the orders.[42]

Conclusion

This chapter has analysed self-regulation as a concept which has been given content in Australia by the NH & MRC guidelines. Criticism of this system which questions the adequacy of self-regulation regarding IVF and human embryo experimentation has been addressed. The major shortcoming arose from the position of the NH & MRC on the status of the human embryo which is outlined in the IVF note. The IEC concept was also examined.

Although IECs are not useless in the NH & MRC model, they face many problems which need to be addressed. The search for an acceptable mode of regulation for embryo research is not yet over.

Notes

1 Baram, M., *Alternatives to Regulation: Managing Risks to Health, Safety, and the Environment* (Lexington Books, Lexington, Mass., 1982) p. 64.

2 Jaffe, L., 'Law as a system of control', *Daedalus* 98 (1969), 406, 409.

3 See e.g. M. Sonnereich, 'Legal and financial encumbrances to clinical research', *Perspectives in Biology and Medicine* 23 (1980a), S 120 at S 137.

4 NH & MRC's Medical Research Ethics Committee, Submission to the Senate Select Committee on the Human Embryo Experimentation Bill 1985, Senate Hansard Report (hereafter called Senate Hansard Report), 26 February 1986, p. 317.

5 Saunders, D., in his submission, Senate Hansard Report, 23 June 1986, p. 2298, outlined the NH & MRC structure and stated:
      I do not believe in legal change which is not enforceable but that community concern can be adequately alleviated by the demonstrated capacity of the IVF [sic] to act in an ethical fashion and to display Australia-wide self-regulation. [Emphasis added.]

6 See generally NH & MRC, 'The NH & MRC and ethical regulation: a short history', *Australian Health Review* 9(3) (1986), 234.

7 Ibid., 234–5.

8 Ibid., 236.

9 The functions of the MREC were to include keeping ethical guidelines under review as well as monitoring their observance and the activities of the IECs: *The First Report by the NH & MRC Working Party on Ethics*, August 1982, p. 13, para. 7.2.4.

10 Senate Hansard Report, 26 February 1986, p. 306.

11 In Victoria embryo research is governed by the *Infertility (Medical Procedures) Act 1984* and has to be approved by the Standing Review and Advisory Committee.

12 Para. 4 (iii) of the IEC note provides on the function of an IEC:
      to maintain a record of all proposed research projects, so that the following items of information are readily available:
         name of responsible institution
         project identification number
         principal investigator(s)
         short title of project

ethical approval or non-approval with date
date(s) designated for review [Emphasis added.]

13 Medical Research Ethics Committee, *In Vitro Fertilization Centres in Australia: Their Observance of the National Health and Medical Research Council Guidelines*, report to the National Health and Medical Research Council, August 1986 (AGPS, Canberra, 1987), pp. 2–3.

14 Ibid., p. 3.

15 Ibid., pp. 3–4.

16 Ibid., p. 5.

17 See e.g. C. Thompson, 'Legal issues arising from peer review processes', *Australian Health Review* 9(3) (1986), 258, 261.

18 See e.g. Senator Carrick, Senate Hansard Report, 26 February 1986, at pp. 443–50.

19 Saunders, D., ibid., p. 2311.

20 Reproduced in the Senate Hansard Report, 26 February 1986, pp. 606–8.

21 Mason, S., submission, Senate Hansard Report, 23 April 1986, p. 1396.

22 Ibid., pp. 2322–3.

23 Wood, C., ibid., p. 99.

24 Levine, R., 'The impact of institutional review boards on clinical research', *Perspectives in Biology and Medicine* 23 (1980a), S98, S99, citing the *Report on Institutional Review Boards* by the United States National Commission for the Protection of Human Subjects of Biomedical and Behavioral Research, 1978.

25 Harman, F., Senate Hansard Report, 26 February 1986, p. 525.

26 Siedlecky, S., ibid., p. 1021.

27 Ibid., p. 1685.

28 See e.g. R. Koval, Supplementary Submission, ibid., pp. 1607–44 and oral evidence, especially p. 1656.

29 NH & MRC, *First Report by the NH & MRC Working Party on Ethics*, para. 7.1.4.

30 Lovell, R., Senate Hansard Report, 26 February 1986, at p. 442.

31 See e.g. ibid., at pp. 395–9 and 443–4.

32 *The Times*, 7 May 1987 and 12 June 1987.

33 See New South Wales Law Reform Commission, *Surrogate Motherhood* (LRC 60) 1988, p. 20.

34 See e.g. J. Brown, et al., 'The costs of an institutional review board', *Journal of Medical Education* (1979), 294.

35 See e.g. J. Katz, 'Who is to keep guard over the guards themselves?', *Fertility and Sterility* 23 (8) (1972), 604, 607 ff.

36 See e.g. R. Levine, *Ethics and Regulation of Clinical Research* (Urban & Schwarzenberg, Baltimore, 1981), p. 233. This criticism is not necessarily valid. The court system is hierarchical but this does mean that lower courts lack, or have diminished, authority.

37 Levine, op cit., n. 24 and S99.

38 Jansen, R., 'The clinical impact of in-vitro fertilization: Part 2, regulation, money and research', *Medical Journal of Australia* 146 (1987), 362, 363.

39 Baram, op. cit, p. 64.

40 Trounson, A. and Wood, C., 'In-vitro fertilization', *Medical Journal of Australia* 146 (1987), 338, 339.

41 Ethical Committee, University College Hospital, London, 'Experience at a clinical research ethical review committee', *British Medical Journal* 283 (1981), 1312. See too J. Faccini et al., 'European Ethical Review Committee: the experience of an international ethics committee reviewing protocols for drug trials', *British Medical Journal* 289 (1984), 1052, 1053. This committee keeps approved projects under review by continuous correspondence with the project physician who informs the committee of any developments which may lead to the stoppage of the project. The NH & MRC's IEC Note (Supplementary Note 1) provides that the IEC should 'keep the progress of research projects under surveillance so as to be satisfied that they continue to conform with approved ethical standards' (Para (i) (b)). It is not clear how this function has been performed.

42 Lewis, P., 'The drawbacks of research ethics committees', *Journal of Medical Ethics* 8 (1982), 61, 61–2. See too D. Weatherall, 'Commentary', 8 (1982) ibid., 63, where he comments on the problems raised by Lewis.

16 | Public policy in a pluralist society

R.M. HARE

What is the proper relation between the moral principles that should govern public policy, including legislation, and moral principles which may be held—often passionately—by individuals, including individual legislators? The adherents of such 'personal' principles often object that proposed laws would allow people, or even compel them, to transgress the principles. Obvious examples are homosexuality and abortion law reform. People who think homosexuality an abominable sin object to the repeal of laws that make it a crime; and those who think that abortion is as wrong as murder of grown people object that a law permitting abortion in certain cases might make it permissible for other people to—as they would say—murder unborn children, or even, if they are nurses and want to keep their jobs, compel them to do so themselves.

So the question we have to consider is really this: what weight ought to be given to the objections of these people when framing and debating legislation and policy? We live in a pluralist society, which means that the moral principles held sacred among different sections of society are divergent and often conflicting; and we live in a democratic society, in which, therefore, policy and legislation have to be decided on by procedures involving voting by all of us or by our representatives; so the question becomes: what attention should we pay, whether we are legislators in parliaments or simply voters in a constituency, to the personal moral opinions of other people, or even to our own? In short, how ought people's moral convictions to affect the actions done by them or by others which influence public policy?

But we cannot address this question until we have answered a prior one, namely: what consideration ought in general to be given to moral principles of any kind when framing legislation and policy? There are three positions on this which I wish to distinguish. The first two seem to me unacceptable, for reasons which I shall give. I will call the first the 'Keep morality out of politics' position. It holds that the function of policy and of legislation is to preserve the interests, which may be purely selfish interests, of the governed; if moral considerations seem to conflict with this function, they should be ignored. Politicians have a moral duty to subordinate, in their political actions, all *other* moral duties to that of preserving the interests of the governed. This position leaves them with just one moral duty; it treats the situation of a government in power as analogous to that of an agent (say a lawyer who might be thought by some to have a duty to preserve the interests of his client at the cost of ignoring some other supposed moral duties).

The difficulty with this position is that no reason is given by its advocates why that should be the politician's supreme and only duty. When a moral question is in dispute, as this one certainly is, we need some method, other than appeals to the convictions of those who maintain the position, of deciding whether to believe them or not. I shall, therefore, postpone discussion of this position until we are in possession of such a method, which will be after we have examined the third position.

The second position goes to the opposite extreme from the first. It holds that morality does apply to political actions and to legislation (very much so). The way it applies is this: there are perfect laws (laid up in heaven, as it were), to which all human laws ought morally to be made to conform. There are various versions of this position, which I shall call generically the 'natural law' position, although that expression also has other different uses. One version says that there is a moral law and that the function of ordinary positive laws is to copy this and add appropriate penalties and sanctions. Thus murder is wrong according to the moral law, and the function of positive law and the duty of the legislators is to make it illegal and impose a penalty.

According to this version all sins ought to be made crimes. But there is a less extreme version according to which not all ought to be: there are some actions which are morally wrong but which the law ought not to intervene to punish. For example, in many societies adultery is held to be wrong but is not a criminal offence. And it may be further added that not all crimes have to be sins. If the law requires people always to carry an identity card, I may be subject to penalties if I do not, but many people would not want to say that I am *morally* at fault. But it does not follow from my not being morally at fault if I break the law that the legislators were morally at fault when they made the law. They might have had very good reasons—even good moral reasons—for making the law (for example, that it would facilitate the apprehension of criminals). A distinction is thus made between what are called *mala in se* (acts wrong in themselves) and *mala prohibita* (acts wrong only because they have been made illegal).

The trouble with this position is very similar to that with the first and extreme opposite position. No way has been given of telling what is in the natural law, nor of telling what sins ought to be made crimes, and which crimes are also sins. That is why, when appeal is made to the natural law, or to a moral law to which positive laws morally ought to be made to conform, people disagree so radically about what in particular this requires legislators to do. To quote the great Danish jurist Alf Ross, 'like a harlot, the natural law is at the disposal of everyone'.[1] Here again we shall have to postpone discussion until we have a safe method of handling such questions as 'How do we decide rationally, and not merely by appeal to prejudices dignified by the name of "deep moral convictions", what legislators morally ought to do?'.

There is another, more serious, thing wrong with the second or natural law position. It assumes without argument that the only moral reason for passing laws is that they conform to the natural law. But there can be many reasons other than this why laws ought to be passed. If, for example, it is being debated whether the speed limit on freeways ought to be raised or lowered, the argument is not about whether it is in accordance with the moral or natural law that people should drive no faster than a certain speed. There may, certainly, be moral reasons, irrespective of any law, why people ought not to drive faster than, or slower than, a certain speed on certain parts of the freeways at certain times and under certain traffic and weather conditions. But that is not what legislators talk about. They talk about what the *consequences* of having a certain law would be. For example, they ask what effect a lower limit would have on the overall consumption or conservation of fuel; what the effect would be in total on the accident figures; whether a higher limit would make it necessary to adopt a higher and, therefore, more costly specification for the design of freeways; whether a lower limit would lead to widespread disregard of and perhaps contempt for the law; and so on. What they are asking, as responsible legislators, is not whether 100 or 110 kph conforms to the natural law, but what they would be doing, i.e. bringing about, if they passed a certain law.

And this requirement to consider what one is doing does not apply only to the decisions of legislators. What I have said responsible legislators do is what all responsible agents have to do if they are to act morally. To act is to do something, and the morality of the act depends on what one is doing. And what one is doing is bringing about certain changes in the events that would otherwise have taken place—altering the history of the universe in a certain respect. For example, if in pulling the trigger I would be causing someone's death, that is a different act from what it would be if I pointed my gun at the ground; and the difference is morally relevant. The difference in the morality of the acts is due to a difference in what I would be causing to happen if I tightened my finger on the trigger. This does not imply that I am responsible for *all* the consequences of my bodily movements. There are well canvassed exceptions (accident, mistake, unavoidable ignorance, etc.), and there are many consequences of my bodily movements that I cannot know of and should not try, such as the displacement of particular molecules of air. Only some, not all, of the consequences are morally relevant.[2] But when allowance has been made for all this, what I am judged on morally is what I bring about.

It is sometimes held that we are only condemned for doing something when we *intend* to do it. This is right, properly understood. If we are judging the moral character of an agent, only what he does intentionally is relevant. But it is wrong to think that we can circumscribe intentions too narrowly for this purpose. There is a distinction, important for some purposes, between direct and oblique intentions. To intend some consequence directly one has to desire it. To intend it obliquely one has only to foresee it. But in the present context it is important that oblique intentions as well as direct intentions are relevant to the morality of actions. We have the duty to avoid bringing about consequences that we ought not to bring about, even if we do not desire those consequences in themselves, provided only that we know that they will be consequences. I am to blame if I knowingly bring about someone's death in the course of some plan of mine, even if I do not desire his death in itself—that is, even if I intend the death only obliquely and not directly. As we shall see, this is very relevant to the decisions of legislators (many of whose intentions are oblique), in that they have a duty to consider consequences of their legislation that they can foresee, and not merely those that they desire.

The legislators are to be judged morally on what they are doing (i.e. bringing about) by passing their laws. They will be condemned morally, in the speed limit example, if they make the limit so high that the accident rate goes up significantly, or so low that it is universally disregarded and unenforceable and, as a result, the law is brought into disrespect. And this brings me to my third possible position on the question of how morality applies to law-making. I will call it the 'consequentialist' position. It says that legislators, if they want to make their acts as legislators conform to morality (that is, to pass the laws they morally ought to pass and not those they ought not), they should look at what they would be doing if they passed them or threw them out. And this means what changes in society or in its environment they would be bringing about.

What legislators are doing, or trying to do, is to bring about a certain state of society rather than some other, so far as the law can effect this— that is, a state of society in which certain sorts of things happen. The legislators are not going themselves to be doing any of these things directly, though, as I have been maintaining, they will be bringing it about intentionally that the things happen, and the bringing about is an act of theirs. There is a school of casuistry which holds that we are not to be held responsible for things which other people do as a result of what we do. I do not think that this school can have anything to say about the question we are considering. For *everything* that happens as a result of the laws that the legislators pass is something that other people do. In the narrow sense in which these casuists use the word 'do' the legislators do nothing except pass the laws. So on this view the legislators are simply not to be held responsible for anything that happens in society; so far as morality goes they can do as they please. I shall, therefore, say no more about this school of casuistry.

Consequentialism as a theory in moral philosophy, which I have been advocating, has received a lot of hostile criticism in recent years. This is because people have not understood what the consequentialist position is. I do not see how anybody could deny the position I have just outlined, because to deny it is to deny that what we are judged morally for (what we are responsible for) is our actions, i.e. what we bring about. What makes people look askance at what they call consequentialism is the thought that it might lead people to seek good consequences at the cost of doing what is morally wrong—as it is said, to do evil that good may come. But this is a misunderstanding. It would indeed be possible to bring about certain desirable consequences at the cost of bringing about certain other consequences which we ought not to bring about. But if the whole of the consequences of our actions (what in sum we do) were what we ought to do, then we must have acted rightly, all things considered.

There is also a further misconception. People sometimes speak as if there were a line to be drawn between an action in itself and the consequences of the action. I am not saying that according to some ways of speaking such a line cannot be drawn; but only that it is not going to divide the morally relevant from the morally irrelevant. In the gun example, nobody would wish to say that my victim's death, which is an intended consequence of my pulling the trigger, is irrelevant to the morality of my act, and that only the movement of my finger is relevant. I intend both, and, as I have said, what I intend obliquely is relevant to the morality of the act as well as what I intend directly. There are a lot of questions, interesting to philosophers of action, which could be gone into here; but I have said enough for the purposes of the present argument.

It is now time to look again at the first two positions I distinguished. What is wrong with both of them is that they ignore what the third position rightly takes into account, namely the consequences of legislation and policy, that is, what the legislators and policy-makers are *doing* by their actions. In short, both these positions encourage irresponsibility in governments. The first position, indeed, does impose on governments a moral duty of responsibility so far as the interests of their subjects go. But what about the effects of their actions on the rest of the world? Ought the British government not to have thought about the interests of Australians when it arranged its notorious atomic tests at Maralinga? Ought it not now to think about acid rain in Norway when regulating power station emissions? Hitler, perhaps, was a good disciple of this position when he thought just about the interests of Germans and said 'Damn the rest'. If we are speaking of moral duties, surely governments have duties to people in other countries. What these duties are, and how they are to be reconciled with duties to the governments' own citizens, is a subject that is fortunately outside the scope of this chapter. I shall consider only what duties governments and legislators have in relation to the states of their own societies which they are by their actions bringing about.

What I have called the natural law position is even more open to the charge of irresponsibility. It says that there are model laws laid up in heaven, and that the legislators have a duty to write these into the positive law of the land no matter what the consequences may be for those who have to live under them. This might not be so bad if we had any way of knowing what was in the model code. But we have not; all we have is a diversity of moral convictions, differing wildly from one another, without any reasons being given by those who hold them why we should agree with them. That is one of the facts of life in a pluralist society. So what happens in practice is that people set up pressure groups (churches are ready-made pressure groups, and there are others on both sides of most disputes), and produce rhetoric and propaganda in attempts to bounce the legislators into adopting their point of view. It cannot be denied that in the course of this exercise useful arguments may be produced on both sides. But when the legislators come to their own task, which is to decide what they morally ought to do, we could wish that they had more to go on than a lot of conflicting propaganda. There ought to be a way in which they can think about such matters rationally, and decide for themselves what they really ought to do. This is especially to be hoped for when they are deciding about embryo experimentation.

Commissions and committees that are set up to help governments decide such questions are often no help at all. In Britain we had a committee chaired by Baroness Warnock which exemplifies what I have been saying.[3] What the Warnock Committee ought to have been doing was to go over the reasons, such as they are, adduced on both sides, and decide which of them are good reasons, and say why. Then they might have been able to sort the matter out, and tell the government and the public where the balance of good reasons lay.

Other government committees in Britain have succeeded in doing this: I could quote the Williams and Wolfenden committees.[4] But the Warnock Committee on Human Fertilization and Embryology almost completely shirked the task of giving reasons for its conclusions, and so left the government and the public in the same position as they were in before: that of having to make up their minds on these complicated, tormenting and crucial questions without much guidance on the reasons for taking one course rather than another. I am not denying that many of the recommendations of the Warnock Committee were quite sensible—only complaining that it gave no indications of *why* they were sensible. I have discussed the committee's way of proceeding in more detail elsewhere,[5] and I shall shortly be dealing with the substance of one of the problems it reported on.

There are two explanations of why the Warnock Committee failed in this way. The first is the one often repeated by Mary Warnock herself. It is that it was much easier to get her committee to agree on its recommendations than on the reasons for them. This is a familiar problem in committees; but all the same I believe that she ought to have tried harder, if they were going to be of help to the government and the public. Other committees have succeeded in doing it, including those mentioned above. The second explanation gives the reason why she did not try harder.[6] She herself is a philosopher, and has a position in moral philosophy which got in the way. She believes that the unthinking reactions of 'outrage and shock' (the phrase is Sir Stuart Hampshire's[7]) which people evince are data on which we can base moral conclusions. That is, if we find people reacting with horror to some suggestion, that by itself is a sign that there is something morally wrong with it. If one takes this view, one is absolved from giving reasons for one's moral opinions. All one has to do is to go round collecting the moral reactions of people, find some way of securing maximal agreement between them, and recommend accordingly. So what Mary Warnock did was to find out what the reactions of the members of her committee were to various proposals (never mind about the reasons for them), and try to get them to compromise with one another so as to produce a unanimous report. She did not succeed, because there are two expressions of dissent in the report; but she did fairly well if that was the limit of her ambition.

It is obvious that this will not do as a means of arriving at rational guidance for governments on moral questions affecting policy. Suppose we were to try this method in a committee in, say, South Africa. A lot of people in South Africa think it is immoral for blacks to swim even in the same private pool, let alone on the same public beach, as whites. So, if a lot of such people found themselves on a government committee about racial policy with somebody of Mary Warnock's philosophical views presiding, the committee would certainly and unanimously recommend the retention of 'whites only' beaches, and would not think it necessary to give any but the most perfunctory reasons. This is simply a recipe for the perpetuation of prejudices without having to justify them. And it is what has happened frequently on committees about IVF, surrogacy, embryo experimentation and the like.

All who handle or advise on such questions ought to be looking for arguments and testing them. So the next thing we need to ask is: what makes an argument on this sort of topic a good one? The answer has been anticipated in what I have said already. Reasoning about moral questions should start by asking what we would be doing if we followed a certain proposal. And what we would be doing is bringing about certain consequences. So what we have to ask first is: what consequences would we be bringing about if we followed it? That is what any responsible government, and any responsible committee advising a government, has to ask first.

It is not, of course, the last thing that they have to ask. They have then to go on to ask which of these consequences are ones that they morally ought to be trying to bring about and which not. But at least they will have made a good start if they have tried to find out what the consequences would be. The question of embryo experimentation illustrates this very well. Suppose that Australian legislators are persuaded by one of the pressure groups in this field, or by the Tate Committee,[8] that they ought to ban all embryo experimentation, and proceed to do so. One consequence is likely to be that the advance of technology in this field will be retarded by the cessation of such experiments, at least in Australia. Perhaps the scientists will get jobs elsewhere in order to continue their experiments; but perhaps, if other governments are taking or likely to take the same line, they will find it difficult to do so. So there will be the further result that the benefits that could come from the research (for example, help to infertile couples to have children, or the elimination of some crippling hereditary diseases) will not be realized.

I have given the strongest arguments on one side. What are those on the other? We might start by thinking that a further consequence of the legislation will be that a lot of embryos will survive which otherwise would not have survived (assuming for simplicity that, as is probably the case, nearly all experiments on embryos using present techniques involve the subsequent death of the embryo). But actually that is wrong: the consequence will not be that embryos survive—at least not in all cases. Whether it is will depend on whether the embryos in question are so-called 'spare' embryos, or embryos created specially for experimentation. If they are spare embryos, indeed, the consequence of the legislation will be that they will survive *the threat of experimentation*. But if we ask what will happen to them if they survive this threat, the answer will be that either a home will be found for them and they will be implanted (perhaps after a period in the freezer) or they will perish, because there is nothing else that can be done with them. In the first case they were not really *spare* embryos. But if we assume that there are going to be at least some embryos which really are spare, and for which, therefore, no home can be found, the result of the legislation will not be different from what it would have been if there had been no legislation: they will perish just the same.

Suppose, however, that, faced with this argument, the legislators were to tighten up the law and say that *no* embryos were to be allowed, or caused, to perish. The consequence of this would be that no embryos would be produced artificially except those for which it was certain that a home could be found. For under such a law nobody is going to produce embryos knowing that no homes can be found for them and that, therefore, he or she will end up in court. So the consequence of this tighter law will be that those embryos will not be produced in the first place.

The same is true of embryos produced especially for research. The result of a ban on such research will be that embryos for whom a home is in prospect will be produced and implanted, but that those (whether spare or specially created for experimentation) for whom there is no hope of implantation will simply not be produced at all. The legislators, in making the decision whether or not to impose such a ban on experimentation, are in effect deciding whether to make it the case that these embryos perish, or to make it the case that they never come into existence at all.

It is not in point here to argue whether the intentions of the experi-menter (to kill or to let die, or not to produce in the first place, for example) make a difference to the morality of his or her actions, or to our assessment of his or her moral character. That is not what we are talking about. We are talking about the morality of the *legislators'* actions, and possibly also about *their* moral character (though it is not clear whether the latter should concern us—the fact that the moral character of some legislators is past praying for does not affect the morality of the legislation they vote for). We are talking about the morality of the actions of the legislator, not of the experimenter, and it is the consequences (the intended consequences) of the legislation that affect this. And since it is the consequences to the embryo and to the grown person that the embryo might turn into which are thought to be relevant here, and the legislation makes no significant difference to these, I shall be arguing that such considerations provide no argument for banning the experimentation: no argument to set against the arguments for allowing it that I have already mentioned.

Suppose, then, that we try to look at the question from the embryo's point of view (though actually, as we shall see, the embryo does not *have* a point of view, and this is important). The alternatives for the embryo are two: never to have existed, and to perish. I cannot see that, if we take the liberty of allowing the embryo a point of view, the embryo will find anything to choose between these two alternatives, because in any case embryos know nothing about what happens to them. So the legislation makes no difference to the embryo. For an ordinary grown human being, by contrast, there is a big difference between never having existed and perishing, because perishing is usually an unpleasant and often a painful process, and frustrates desires for what we might have done if we had not perished. But for the embryo it is not unpleasant to perish, and it has no desires.

I conclude that from the point of view of these embryos (namely those for whom the alternatives are as I have described) the legislation makes no difference. But now what about the point of view of the grown person that the embryo might develop into if it were implanted? That grown person certainly would have a point of view; he or she would have desires and would not want to perish now after having grown up. But is there any difference, for this grown person, between not having been produced as an embryo in the first place and, after having been produced, perishing before achieving sentience? I cannot see any; so it is hard to avoid the conclusion that the legislation makes no difference to the grown person either. The Tate Committee, rightly in my opinion, attached great importance to the potential that the embryo has of becoming a grown person.[9] But it drew what seems to me the wrong conclusion from this potential. What makes the potential of the embryo important is that if it is not realized, or is frustrated, there will not be that grown person. But if, as in the cases we are considering, there will not be that grown person anyway, how is the potential important? Indeed, *is* there really any potential? That is, if what is important is the possibility (this word is to be preferred to 'potential') of producing that grown person, and there is no such possibility (because the legislators have a choice between either doing something that will result in the embryo that would develop into that grown person not existing, or doing something that will result in it perishing) it looks as if the legislators can forget about *this* reason for imposing a ban on experimentation. I could go on to say just why the Tate Committee was led into this false move; but instead I will refer to Stephen Buckle's chapter on the subject.[10]

It seems, therefore, as if the reason we have been considering in favour of a ban, namely that it is necessary in order to save the lives of embryos, falls down; for this is only a reason if thereby the possibility of there being those grown people is preserved, and in these cases this is not so. They are analogous to a case in which the embryo has a defect because of which it is sure to perish before it develops into a baby; in that case is there any moral reason for preserving it now?[11] They are also analogous to the case where because of 'cleavage arrest'[12] there is no hope of the embryo ever becoming a child; such embryos just stop developing, and in the present state of *in vitro* technology nothing can be done to start development again. It is hard to see what is lost if such embryos with no potentiality for turning into babies are destroyed, since they will perish anyway; and it is just as hard to see why the same does not apply to other embryos with no hope of survival.

But the preservation of the embryo is the main—indeed the only significant—reason given for imposing a ban; and since, so far as I can see, the reasons given on the opposite side, also concerned with the consequences of imposing it, are much more cogent, and affect many more lives much more powerfully (the lives of those who will be given children if the experiment leads to advances in techniques, the lives of those children themselves, and the lives of those who will otherwise suffer from genetic defects which research could help eliminate), I conclude that rational and responsible legislators would not impose a ban, and that clear-headed committees who could tell the difference between a good and a bad argument would not recommend it. I give this as an illustration of how those concerned with such questions should reason about them. I should have liked to do the same for other questions in this field, such as surrogacy, IVF by donor, and indeed the whole question of whether all artificial methods of reproduction should be banned, as the Vatican seems to think. But I have had to be content with an illustration.

To put the matter bluntly: we should stop wasting our breath on the question of when human life begins. Even if we grant for the sake of argument that it begins at fertilization (however that is defined)—even if we grant that there is a continuity of individual human existence from that time, so that I can answer Professor Anscombe's strikingly phrased question 'Were you a zygote?' in the affirmative[13]—it is going to make no difference to the moral question of what the law ought to be on embryo experimentation. For imagine that I am a grown person who was once a zygote produced by IVF. In that case, I am certainly very glad that I was produced, and that nobody destroyed me, or for that matter the gametes that turned into me. But if you ask me whether I wish there had been a law at that time forbidding embryo experimentation, I answer that I am glad there was no such law. For if there had been, then very likely the IVF procedure which produced me would never have been invented. And such a law could not, in principle, have done anything for me. For though, if I had come into existence, it would have prevented my being destroyed, it would also have made false the antecedent of this hypothetical: I never would have come into existence in the first place.

I have spoken generally throughout of the embryo and the grown person, and not mentioned much the stages in between, such as the fetus, the neonate and the child. This is because the point I have been trying to make can be made clearly for the two extreme cases. What implications all this has for neonates and fetuses—and, for that matter, for pairs of gametes before fertilization—requires further discussion. But it is a big step towards clarity if we can see that at any rate in the case of the embryo it simply does not matter morally whether, or at what point in time, it became an individual human being. What matters is what we are doing to the person, in the ordinary sense of 'person', that the embryo will or may turn into.

Notes

1   Ross, A., *On Law and Justice (Om ret og retfaerdighed)*, trans. M. Dutton (London, Stevens, 1958).

2   See my *Moral Thinking* (Oxford University Press, Oxford, 1981), pp. 62 ff and references.

3   *Report of the Committee of Inquiry into Human Fertilization and Embryology* (Mary Warnock, chair), (HMSO, London, cmnd 9314, 1984). A much more helpful report on the same subject, for the European Commission, is J.C.B. Glover, *Fertility and the Family* (Fourth Estate, London, 1989).

4   *Report of Committee on Obscenity and Film Censorship* (B.A.O. Williams, chairman), (HMSO, London, cmnd 7772, 1980); *Report of Committee on Homosexual Offences* (Sir John Wolfenden, chairman), (HMSO, London, cmnd 247, 1957).

5 See my '*In vitro* fertilization and the Warnock Report' in R. Chadwick (ed.), *Ethics, Reproduction and Genetic Control* (Croom Helm, London, 1987).

6 See my 'An ambiguity in Warnock', *Bioethics* 1 (1987).

7 Hampshire, S., 'Morality and pessimism' (1972) reprinted in his *Public and Private Morality* (Cambridge University Press, Cambridge, 1978).

8 Senate Select Committee on the Human Embryo Experimentation Bill 1985, *Human Embryo Experimentation in Australia* (Senator Michael Tate, chairman), (AGPS, Canberra, 1986).

9 Ibid., pp. 8, 25.

10 Buckle, S., 'Arguing from potential', Chapter 9 of this book.

11 See my 'A Kantian approach to abortion' in M. Bayles and K. Henley (eds), *Right Conduct: Theories and Applications*, 2nd ed. (Random House, New York, 1989).

12 See K. Dawson, 'Segmentation and moral status: a scientific perspective', Chapter 6 of this book.

13 Anscombe, G.E.M., 'Were you a zygote?' in A.P. Griffiths (ed.), *Philosophy and Practice*, Royal Institute of Philosophy Lectures 19, supp. to *Philosophy* 59 (Cambridge University Press, Cambridge, 1985).

17 | Biological processes and moral events

STEPHEN BUCKLE

In the course of discussing the moral acceptability of embryo research, the Warnock Report makes the following observation:

> While . . . the timing of the different stages of development [of the embryo] is critical, once the process has begun, there is no particular part of the developmental process that is more important than another; all are part of a continuous process, and unless each stage takes place normally, at the correct time, and in the correct sequence, further development will cease. Thus biologically there is no one single identifiable stage in the development of the embryo beyond which the in vitro embryo should not be kept alive. However, we agreed that this was an area in which some precise decision must be taken, in order to allay public anxiety.[1]

This passage is important because it draws attention to a way of thinking about the moral issues raised by the new biotechnology which has a considerable appeal. The way of thinking in question sees a crucial moral significance in the continuity of the processes which constitute the development of a human infant. It understands the continuity of the developmental processes to debar the drawing of any moral distinctions except at the beginning and end of the process. The view is commonly, though not always, expressed in the claim that the drawing of moral distinctions in such cases is wholly *arbitrary*.

In the quotation, the Warnock Report appears to concede the importance of this way of thinking, only to deny it in the final sentence. For, if there is no biological stage at which a crucial moral difference is generated regarding how to treat *in vitro* (or other) embryos, how is a precise decision to help? Rather than allaying public anxieties, will not such a decision *provoke* anxieties by its apparent arbitrariness? For if the decision is not based on a biological difference, on what could it be based?

The purpose of this chapter is to argue that the Warnock Report's approach is indeed justifiable, despite the popularity of the kinds of objections just provided. It will be argued, that is, that the continuity of biological processes is no bar to making precise and non-arbitrary moral decisions about the treatment of human embryos. It is important to show this not least because of some politically influential treatments of such issues which follow the popular line. In at least one such case, the above-quoted passage from the Warnock Report is employed to *support* the popular view.

The case in question is the majority report of the Australian Senate Select Committee on the moral and legal issues raised by the prospect of embryo experimentation, *Human Embryo Experimentation in Australia* (the Tate Report).[2] The Tate Report argues against allowing any destructive non-therapeutic experimentation on human embryos, no matter what their stage of development, by stressing that, after fertilization, there is 'a continuum of development until birth' (para. 3.21). In particular, it argues that there is no 'marker event' (i.e., no event which introduces a change in the moral status of the embryo) after fertilization and before birth.[3] To support this view, it lays particular stress on the Warnock Report's claim that 'once the process has begun, there is no particular part of the developmental process that is more important than another; all are part of a continuous process' (quoted at para. 3.17).[4]

It is possible to interpret arguments of this kind in two ways. The more common interpretation is that already referred to—that where we are dealing with continuous processes, any form of demarcation, or line-drawing, is simply arbitrary. The second interpretation is stronger: it holds that the continuity of the biological processes must exclude marker events, because processes are fundamentally different from events, and necessarily exclude them in all cases. Therefore they must exclude them in this particular case. (A picture which may encourage this view is the idea that processes 'flow' whereas events do not: so processes exclude events because processes are smooth and continuous, whereas events are *irruptions*, breaks in what otherwise would be smooth and continuous processes.) This second interpretation is rarely stated explicitly, but often seems to be used to buttress the first against criticism. Since it implies that processes and events *must* be sharply distinct, any problems encountered by views such as the first interpretation can be regarded merely as problems of *formulation* and not of substance. However, it will be argued in this paper that, on either interpretation, this kind of argument is not successful. To show this, it will be best to examine the second interpretation first, in order to avoid complications generated by its influence on arguments which explicitly depend on the first interpretation.

It is certainly true that the process of development from fertilized egg to human baby is a continuous process, but from this fact it does not follow

that markers are impossible. This is because processes, no matter how continuous, are not fundamentally different from events. What distinguishes
an event from a process depends on the context of inquiry—they are not,
as the second interpretation seems to suppose, different kinds of natural
phenomena. One way of showing this is by considering the same occurrence
in the light of different time-scales: for example, kicking a ball. If I kick a
ball, this is certainly an event, something which happens at a particular
time. If we employed a slow-motion camera, however, we would not be able
to identify a precise moment (a specific frame on the camera film, for
example) when the ball could be said to be kicked, for what had seemed
an instantaneous event now is seen to be a process which occurs over time—
first the foot makes contact with the outer surface of the ball; the ball distorts
as the foot invades the space previously occupied by the ball alone; as the
foot continues on its path we see the ball begin to return to its original
shape until, as it regains its original shape, it begins to lose contact with
the foot, and then follow an independent path.

In considering this process, no precise moment presents itself as *the*
moment when the ball is kicked. But, despite any initial surprise that this
should be so, there is no deep problem here. This can be shown in the
following way. If we ask 'Exactly *when* was the ball kicked?' what should we
answer? When the foot made contact? When the foot and ball were in the
most intimate contact (the point of maximum distortion of the ball)? Or
when the ball left the foot? These questions appear difficult to answer because
they prompt us to look for a precise moment when the event occurred, and
such a moment is not clearly revealed by the camera. However, in so
prompting us the questions lead us astray. The difficulty we have in answering
them is not testimony to a difficult problem, but to having been misled by
our own original question. The only adequate answer to that original question
is to insist that the *whole process* caught by the camera is the event of kicking
the ball. Kicking a ball is not only an event which occurs at a particular
time, it is also a process which occurs over time. This may seem paradoxical,
but only if we fail to recognize that in speaking thus we are employing two
different time-scales. The difference between processes and events is not a
difference between two fundamentally different kinds of thing, but (in this
sort of case, at least) between different time-scales, time-scales which reflect
the different interests underlying our different forms of inquiry, and which
are presupposed by the correspondingly different forms of language we employ
for different tasks.

If we now turn to consider the case of human fertilization, we find
that the same principles apply. The fertilization of an egg by a sperm is the
event that begins the process of development which culminates in the birth
of a human baby. So significant has this particular event seemed to many
(including the majority of the Australian Senate Committee) that it is
defended by them not only as a suitable marker event for the recognition
of moral and legal status, but also as the only possible such event. However,

if we examine this event carefully, we find that our increased knowledge of human conception has put us in a position not unlike that encountered in the case of the slow-motion camera. It is no longer obvious what fertilization is (that is, what the scope of the concept is), because our use of the word 'fertilization' is not pre-adapted to discriminate between cases discernible only with the aid of modern medical technology. In its ordinary employment, 'fertilization' refers to the egg and sperm getting together, becoming intimately acquainted as it were, so that the developmental processes get underway. Does this mean, then, that fertilization has occurred once the sperm is inside the egg? Or not until syngamy? Or at some intermediate point? The matter cannot be settled by identifying some instantaneous event, not least because there are none (for example, several hours elapse before the sperm is completely inside the egg). The best solution seems to be to insist, once again, that what is an event from one standpoint is, or can be, a process from another; and thus to conclude that fertilization is that causal sequence beginning when the sperm and egg first interact, and ending when syngamy is complete. But to conclude thus is, in the first place, to propose a convention to govern the proper employment of a term in our language, not to insist on an obvious matter of fact; and, in the second, to allow that 'fertilization' is the name for a biological *process*. But this no more implies that fertilization cannot be a marker event than it is true that causal processes cannot constitute an event. Fertilization can be understood to be a process, and is even perhaps best understood as a process; but this in no way precludes its being a marker event. The distinction between processes and events is contextually sensitive, and much confusion arises when this is not recognized.

The second interpretation of the argument from biological continuity, that the very fact of continuity necessarily precludes marker events because events and processes are different in kind, therefore fails. It is no argument against the possibility of marker events that there is a continuous process from the first interaction of sperm and egg to the birth of the baby. Whether markers can or cannot be discerned depends not on the continuity or otherwise of the biological processes, but on whether the achievement of a certain level of development, however it comes about and whatever it leads to, makes some kind of morally significant difference. A morally significant difference arises at least in those cases where a consideration of moral significance comes to bear. What these considerations are has not been directly addressed here, but candidates are not hard to find: for example, sentience, viability, or (of course) potentiality. In this light, it might even seem that a rather brisk conclusion is warranted, thus: regard for considerations such as these will help to resolve moral problems; fussing about continuous processes will not.

Such a brisk conclusion is not warranted, however, because the argument so far has failed to address the first interpretation of the continuous processes argument. It will be remembered that that interpretation does not depend on any sharp distinction between processes and events. It simply holds that, where we are dealing with continuous processes, all demarcations are quite arbitrary. It may, therefore, even be understood as a critique of the reply given above to the second interpretation. This can be illustrated as follows. It is all very well (it might be said) to identify the achievement of certain stages of development as morally significant markers, but the very continuity of the processes in question here mean that this is an achievement of little, if any, moral use. Morality is a practical matter, so our deliberations on an issue are morally valueless if they do not help us to decide how to act on the matters that concern us. Vague talk about stages of development, of processes constituting events, etc, does not help us to decide where to draw lines governing our research practices. For that task we need markers as precise and exact as our technological expertise itself. But our expert knowledge shows only continuities. There are thus no natural markers revealed in our research, so establishing a marker, for whatever reason, will be an arbitrary decision.

Putting the matter this way helps to show the strength of the first interpretation: by avoiding the second interpretation's conceptual muddle about the relationship between processes and events, it is able to focus on the crucial practical matter. However, it is also, for that very reason, comparatively easy to meet. Drawing a precise line where there is no correspondingly precise natural marker means that the line in question could have been drawn somewhere else. It does not, however, mean that the line could have been drawn *anywhere* else. There may be several possible answers to the question of what counts as fertilization, for example, but it does not follow that we could say anything. Whatever the initial imprecision of such a term, some answers to the question can be readily ruled out as unreasonable or even absurd. Whatever else it is, fertilization is what happens to the egg to get the developmental process underway; and this fact functions as a limit on what answers to the question are reasonable. Applying this more generally, we can say that drawing a precise line for legal or moral purposes in cases where there is no precise natural marker to follow is not arbitrary where it is not unreasonable. As long as it is in the right general area to tie up with the relevant moral feature(s), a precise line can quite reasonably be drawn.

Our legal practice already draws such lines in a variety of cases. For example, legal adulthood is attained at the age of 18, despite the fact that this is only a rough indicator of the age at which the appropriate qualities come to fruition—qualities which develop gradually and continuously, and at quite different rates in different individuals or circumstances. But a precise line is needed for legal purposes, and this is generally accepted as a reasonable point at which to draw the line. The same approach can be employed in cases concerning early human development. Exactly where the line is drawn is less important than that a line *be* drawn in the general area of the appearance of the morally relevant feature. A line drawn in the appropriate general area will be a reasonable line; and, *ipso facto*, it will be a non-arbitrary line. Of course, in those cases where it is necessary to be morally scrupulous, or where for a variety of reasons it might be necessary to avoid even the suspicion of doing wrong, some kind of safety margin should be built in, thereby ensuring that we err on the side of caution rather than carelessness. By doing so, interestingly enough, we make the task of drawing a precise boundary all the easier (although we also leave ourselves open to pressure to have the boundaries revised should new circumstances arise or old interests revive). Despite widespread convictions to the contrary, then, it is perfectly possible to settle on precise markers to govern research practices and resolve legal problems.

This does not mean that there is no problem about settling on marker events. In fact, there is a problem here, but it is a problem about moral beliefs. This can be illustrated as follows. It has been argued here that, despite the fact that there is a continuous developmental process from fertilization to birth of a human baby, it is possible to determine marker events nonetheless. But this does not, of course, tell us what markers we should adopt, or even that there are such (morally significant) markers. Both issues require, for their settlement, the application of a moral viewpoint. Different moral views will recommend the adoption of different events as markers, or may even resist thinking in terms of markers in many cases.

The continuity of the developmental processes in the formation and development of the human embryo thus raise no special problems concerning the establishment of marker events for the purposes of regulating research. Rather, the central questions are whether the early embryo should be recognized as an individual worthy of respect in its own right, and whether specific research programs can be justified by reference to their (intended or likely) consequences. Whether the Warnock Report or the Australian Senate Committee's report provide adequate answers to questions such as these is, of course, quite a separate matter.

Notes

1 Warnock, M., *A Question of Life: The Warnock Report on Human Fertilization and Embryology* (Basil Blackwell, Oxford, 1985), p. 65.

2 Senate Select Committee on the Human Embryo Experimentation Bill 1985, *Human Embryo Experimentation in Australia* (Senator Michael Tate, chairman), (AGPS, Canberra, 1986). (Textual references are to paragraph numbers of this report.)

3 This is a simplification of the report's position, but not, I believe, an unwarranted one. The report says only that *as yet* there is no 'compelling evidence' for the existence of a marker event (para. 3.19). However, given its appeal to the continuity of the biological processes, it is not clear what could count *as* evidence

for a marker event. Does the committee envisage the future discovery of a *discontinuity* in embryonic development?

4 In stressing this point the Senate Committee attaches most weight to one point of the quoted passage which is clearly vulnerable to objections. For, while it is certainly true that every stage in the embryo's development is equally necessary for that development to come to fruition, it is not therefore true that every stage is equally important. Morally speaking, the stage at which the brain is formed would seem to count for more than the stage at which the fingers or other external features are formed, even though the brain cannot be formed without the successful prior completion of the earlier stages.

18 | Embryo experimentation: The path and problems of legislation in Victoria

BETH GAZE AND PASCAL KASIMBA

The *Infertility (Medical Procedures) Act 1984* (Victoria) was an unprecedented step towards the regulation of the issues raised by *in vitro* fertilization (IVF). Elsewhere, investigation of the legal, social, ethical and other implications of IVF was either still under way or yet to begin. Even today very few legislatures have put together any form of regulation as comprehensive as the Victorian initiative.

The purpose of this chapter is two-fold. First, we review the developments in Victoria which led to the passage of the *Infertility (Medical Procedures) Act 1984*. We trace the various underlying influences and identify problems in the process of deciding and implementing policy in parliament in an area where strongly held views fundamentally conflict. Second, we focus on embryo experimentation because, more than any other issue, it became the subject of intense contention. Hence, we will examine the provisions of the Act that deal with this key issue to see how legislative control of this aspect of IVF has operated in practice. This study is relevant to any future attempts, in Australia or in any other country, to regulate research relating to new birth technologies or any other scientifically complex and morally controversial field.

The committee stage

The *Infertility (Medical Procedures) Act 1984* was preceded by a committee appointed by the government of Victoria in 1982 to consider the social, ethical and legal issues arising from IVF and chaired by Professor Louis Waller (the Waller Committee). The impetus for this action was the success of Victoria's IVF scientists. Victoria's first IVF baby, only the third in the world, was born in 1980. This raised IVF developments to the headlines, where the process was initially hailed uncritically by the press as a miracle and doubts concerning the associated ethical, social and legal questions received little media attention. With the procedure a *fait accompli* in Victoria, and ever-growing numbers of people entering the program, these questions could not be avoided. Over the following years, doubts and opposition began to receive attention in Victoria, and debate over IVF commenced.

The existence of IVF raised a number of legal problems requiring attention. For example, the use of donor gametes posed problems of defining the legal parenthood of the child born as a result, with implications for maintenance, custody and inheritance law. These implications had already been raised in relation to children born as a result of donor insemination. Other issues specific to IVF went beyond purely legal matters. They included the status to be afforded to the embryo, issues of access, resource allocation for IVF patients and certification of hospitals in which the procedures could be performed. The possibility of the use of techniques such as egg and embryo freezing, and experimentation on human embryos raised ethical questions on which a wide variety of views exist in any pluralist community.

Some of the legal issues required that parliament deal with IVF, at least to some extent, and the ethical issues and role of the government in the health system suggested that this attention should not be restricted solely to dealing with issues affecting children born through IVF. The appointment of a committee to consider these issues was seen as ensuring that

> the government will be able to come to a considered decision on how best to deal with the issues raised by In Vitro Fertilization having taken into account the views and beliefs of all sections of the community as determined by the Committee.[1]

From a purely political perspective, some governments in Australia have not been eager to intervene in areas of moral controversy, and the Australian law of abortion provides a good example of this. Once an issue is raised, one way of avoiding direct political pressures over it is to set up an independent committee. This course was chosen for IVF.

The Waller Committee's terms of reference required it to consider:
• whether the process of *in vitro* fertilization should be conducted in Victoria, and if so, the procedures and guidelines that should be implemented in respect of such processes, by legislation or otherwise;
• whether the community and the parties (including the donors, embryo and technicians) involved in the process of *in vitro* fertilization had any rights and/or obligations and if so, whether such rights and/or obligations should be enforced, by legislation or otherwise. This required consideration of questions of access to and selection for IVF, and the treatment of embryos;

- make recommendations upon such related matters as it considered appropriate;
- co-opt such other persons to assist and advise as it deemed appropriate; and
- make an interim report to the Attorney-General within three months of the committee first convening.

These terms of reference gave the Waller Committee power to expand its inquiry beyond the matters mentioned in the first and second terms of reference. Those terms were fairly general, requiring consideration of whether the practice of IVF involved 'undesirable social and moral practices' or may infringe on the rights of any participant in the process, of problems of resource allocation and patient selection, and of the disposal of embryos created but not used for transfer. The committee was put under pressure to act quickly by the requirement that it produce an interim report within three months of its first meeting.

The Waller Committee had a total of nine members, four women and five men, from varying social and professional backgrounds, including law, teaching, medicine, social work and moral theology. This composition reflected the government's perception that the issues involved required varied perspectives for their resolution, although the power to co-opt advisers was left to the committee itself. Two notable omissions from the committee's composition were a research scientist working in the IVF field, who could ensure the factual accuracy of the committee's deliberations on the laboratory, as opposed to the clinical, aspects of IVF, and a moral philosopher. Although the committee heard presentations from experts in both these fields, its reports were drafted and settled without their assistance.

The Waller Committee faced a difficult task. Formally established two months before the Warnock Committee in the UK, it was required to report on a problem which had until that time received little systematic discussion or analysis in Australia. Parallel developments were also occurring elsewhere in Australia at that time, for example, the creation by the NH & MRC of a working group on medical ethics, later formalized as the Medical Research Ethics Committee. Committees of inquiry into IVF were set up by all the other states, and at the Commonwealth level.[2]

The resources provided to the Waller Committee by the Victorian Government to achieve this task were limited. It had no full-time support staff, but was provided with limited secretarial assistance and library resources by the Law Department and the Law Reform Commissioner. All members of the committee served part time, and research assistance provided was limited. Despite these limitations, the committee produced three reports over two years. Its method of proceeding was generally to invite written submissions from the public by advertisements in the daily papers, and to hear presentations from invited experts on specific areas. In addition, a public hearing was held for the interim report and a discussion paper inviting written submissions issued for the second report on the use of donor gametes in IVF. The Waller Committee's Interim Report was produced in September 1982, followed by the *Report on Donor Gametes in In Vitro Fertilization* in August 1983, and the *Report on the Disposition of Embryos Produced by In Vitro Fertilization*, which also examined embryo experimentation and surrogacy in the context of IVF, in August 1984.

Given that the Waller Committee's job was to advise the government on the desirability of legislation (and its form and content), the most that could be achieved was to reach a broad consensus on a framework to deal with the most important aspects of the issues. Each report includes dissenting statements, and has been criticized by groups opposed to the majority recommendations. The committee saw its role in part as being to determine the acceptability to the Victorian community of various forms of IVF treatment and embryo experimentation.

The Waller Committee's reports

All three of the Waller Committee's reports eventually formed the basis of the recommendations implemented in the *Infertility (Medical Procedures) Act 1984* (Vic.). The main features of each report, particularly as they relate to embryo experimentation, will be mentioned and linked to their scientific and general context to give a better understanding of the development of the committee's approach.[3]

The first report gave some background information on the development of IVF techniques. Discussion of the issues arising from IVF was limited to only the 'most common situation' where a married couple is treated and their own gametes are used. A number of issues requiring further attention were identified in the first report, for example, the use of donor gametes, freezing and storage of embryos, surrogate motherhood, creation of embryos specifically for research purposes, and cloning and other genetic engineering techniques.

The Waller Committee reported that there was both strong support for and substantial opposition to IVF. It considered IVF in the 'most common situation' to be acceptable to the community and recommended that it be recognized as such. It then proceeded to consider what safeguards were necessary, and recommended that it should only be carried out in approved hospitals, that counselling be made available, that informed consent should be obtained from couples undergoing the treatment, and that to reduce the problems faced by infertile couples generally, an education program should be provided in schools and the media aimed at increasing awareness and understanding of the problem of infertility. As a basis for selection to an IVF program, the committee recommended that IVF should be limited to cases where other appropriate means of treatment had been attempted and infertility had proven unresponsive or the couple had sought alternative treatment for a period in excess of 12 months. The continuation of IVF in the most common situation was accepted by the whole committee on con-dition that all fertilized eggs were transferred to the woman's uterus. A majority of six believed that where too many eggs were fertilized for all to be transferred, the wishes of the couple as to disposal of the excess embryos should be respected, and were prepared to allow continuation of embryo freezing until the committee considered it. The minority believed further study was needed on the question of surplus embryos and that procedures identified for later attention such as embryo freezing should not be used in Victoria until they had been considered.

The second report, on the use of donor gametes in IVF, was published before any pregnancy had resulted from the use of donated eggs alone, but after one report of successful transfer of a donated embryo, which later miscarried. The intense public and media interest in these issues is reflected in the introductory section of the report, which describes the controversy over the suspension of the IVF donor gamete program at a Melbourne hospital. The majority report repeated the need for a community education program, this time on infertility and the use of donor gametes in IVF, noting that there had been 'overall, no extended public consideration of the matters which it has had to examine'.[4] It approved the use of donor sperm and eggs as well as donor embryos, again conditional on requirements as to authori-zation of hospitals, counselling and informed consent, this time of both recipient couple and donor. Additional recommendations were made con-cerning maintenance of records of donated genetic material, that only one source be used for each donation of gametes to avoid confusion as to a child's genetic origin, and for legislative changes to deal with the legal status of children born as a result of the use of donor gametes.

Following the second report, two bills were introduced in the Victorian parliament in March 1984. The Status of Children (Amendment) Bill which clarified the position of gamete donors and the status of children born from donated gametes, was passed without delay. The first version of the Infertility (Medical Procedures) Bill[5] dealt with the recommendations made to date by the Waller Committee, directed mainly to regulating the practice of IVF and the use of donor gametes. It did not deal with freezing, storage or disposal of gametes or embryos, nor with embryo experimentation, as these issues were still under consideration. This factor contributed to an adjournment of debate on the Bill until the next session of parliament so that these issues could also be considered. Some provisions of the Bill were slightly different from the committee's recommendations: for example, the pre-condition for IVF treatment that infertility had proven unresponsive to alternative treat-ment or treatment had been sought for at least 12 months became, in the Bill, and subsequently in the Act, a requirement that treatment for infertility had been commenced at least 12 months before entering the IVF program.[6]

Before the Waller Committee's third and final report, *Disposition of Embryos Produced by In Vitro Fertilization*, was released in August 1984, sperm freezing was well established not only for IVF but donor insemination as well. Egg freezing attempts had not yet had any success in producing fertilizable eggs, and embryo freezing, used successfully for many years in veterinary practice, had been introduced in local and overseas IVF clinics. The first pregnancy in the world from a frozen embryo had occurred in Victoria through the Queen Victoria Medical Centre in March 1984, although it ended in miscarriage.[7] In addition, before the report was delivered, the case of the 'Rios embryos' had become a media *cause célèbre*. This case arose when a couple for whom two embryos had been formed *in vitro* and frozen died in an air crash early in 1984, leaving behind the problem of disposal of the embryos. The couple were both American citizens admitted as patients in 1980, and the embryos had been formed from the wife's eggs and donor sperm in November 1981.[8] These areas, which exemplify factors that influenced the Waller Committee's work, were to feature in the final report.

This report dealt with issues such as freezing and storage of embryos, the fate of frozen embryos, embryo experimentation, surrogate motherhood arrangements in IVF, and future supervision of the area of fertility and reproduction. As would be expected on such contentious issues, several members of the Waller Committee dissented on aspects of the recommendations. One member considered both freezing of embryos and creation of embryos or use of surplus embryos for experimentation totally unjustified. Another was prepared to allow freezing and storage of embryos only for medical reasons within IVF treatment requiring the delay of embryo transfer, such as bleeding by the recipient woman. On the question of embryo experimentation, two members from the medical field argued that it was ethically acceptable to create embryos for research purposes. The majority view was that it was unacceptable to create embryos solely for the purpose of experimentation but that as research was essential to the improvement of IVF techniques, excess or spare embryos that could not be transferred to women could be made available for research, provided it was undertaken immediately and with the specific consent of the people providing the genetic material, and the embryo was not frozen for use in some future unspecified research project. The Waller Committee approved of the freezing of sperm and embryos, in the latter case because of the difficulties with egg freezing and because it makes both collection and use of eggs, as well as donation of eggs and embryos, more effective. However, freezing of embryos was regarded as still experimental, and was to be restricted to the shortest possible time, in hospitals specifically authorized for the procedure. Where embryos were stored, the consent to storage should provide for later disposition of the embryos.

The *Infertility (Medical Procedures) Act 1984*

A revised Infertility (Medical Procedures) Bill was debated in parliament in October 1984. Many provisions of the Bill were amended by the opposition in the upper house where the government party was in a minority. Some amendments brought the Act closer to the Waller Committee's recommendations, while others moved it further from them.

In its final form, the Act covered many of the areas included in the reports concerning the clinical application of IVF. These include the approval of hospitals to carry out IVF; keeping records and disclosure of information on donors of gametes, IVF couples and children; and selection, counselling and consent of patients for IVF programs. In relation to research the Act prohibited cloning[9] and cross-species fertilization of human gametes, and freezing of embryos except for the purposes of IVF treatment. The provisions on embryo experimentation have been the most contentious. For this reason, they will be examined separately together with the related provisions on ova and embryo freezing, and the Standing Review and Advisory Committee.

Embryo experimentation

The provisions pertaining to experimentation are mainly contained in s. 6 and s. 29 which creates the Standing Review and Advisory Committee on Infertility (SRACI), a body charged with the duty of administering s. 6(3). Interpretation of s. 6 has been plagued by uncertainty since the Act was passed. At issue has been the nature of the prohibitions and permissions regarding embryo experiments. The problem is exacerbated by lack of adequate definition of major terms, including 'experimentation' and 'IVF human embryo'. Experimentation, for instance, is defined in an indirect way. Section 6(3) provides that 'a person shall not carry out an experimental procedure other than an experimental procedure approved by the Standing Review and Advisory Committee'. 'Experimental procedure' is then defined in s. 6(4) as:

> a procedure that involves carrying out research on an embryo of a kind that would cause damage to the embryo, would make the embryo unfit for implantation or would reduce the prospects of a pregnancy resulting from the implantation of the embryo.

Section 3 defines 'approved experimental procedure' as 'an experimental procedure directly related to the alleviation of infertility and approved by the Standing Review and Advisory Committee'.[10] This provision seems to imply two things: that experiments that do not damage an embryo are permitted while those that do cause damage are permitted only when the approval of the SRACI has been sought and obtained. But there is a problem with the latter point. The Act does not say how far the SRACI can go in granting approval. For instance, can the SRACI approve experiments that involve embryos that have developed beyond 14 days? The Waller Committee recommended that only excess or spare embryos which have not developed beyond 14 days after fertilization be the subject of damaging experiments.[11] Under the rules of interpreting statutes in Victoria, the recommendation can fill the gap in the Act.

This interpretation of the SRACI's powers is, however, not universally shared. Early in 1989 it approved a research proposal involving two-day-old spare embryos. Amid conflicting claims about its power to do so, the SRACI was, in effect, overruled by the minister who 'requested' the researchers to defer the experiments pending a review, by the SRACI, of embryo experimentation.

Ova and embryo freezing

Another anomalous situation in interpreting s. 6 arose from the provisions of subss (5), (7) and (8) of the section. Subsections (7) and (8) permit embryo and ova freezing techniques while subs. (5) prohibits the fertilization of ova except for purposes of implanting them in the womb of a woman. According to researchers involved in developing these techniques, their work could not proceed ethically unless some ova were frozen, thawed, fertilized *in vitro* and observed for abnormalities during the early stages of cell division. In this way there was a conflict between these subsections.

But separate problems arose for subs. (5) that go to the root of employing legislation in an area where knowledge is changing all the time. For instance, when the Waller Committee's reports were made, fertilization was understood as a more or less instantaneous event. Later understanding showed that this was not so. The subsection became the subject of amendment by the *Infertility (Medical Procedures) Amendment Act 1987*. This was precipitated by an application, in 1986, for approval of a research project which involved testing the success of microinjection of a single sperm into an egg before fertilization was completed by the fusion of the two pronuclei from the gametes. The Standing Review and Advisory Committee split equally on whether or not it could be approved under s. 6(5) which had not at that stage been proclaimed to commence operation. Despite legal advice from Victoria's Solicitor-General that s. 6(5) was not operative, the SRACI requested that the problem be resolved by amending the section to clarify the reference to fertilization as a process. The amendment, therefore, allowed the SRACI to approve research up to but not including syngamy, defined as 'the alignment on the mitotic spindle of the chromosomes derived from the pronuclei'.[12] Thus permission of the SRACI is necessary for tests which involve fertilizing eggs before the stage of syngamy is reached.

The Standing Review and Advisory Committee

Under s. 6(3) and s. 9A(1) an 'experimental procedure' cannot be carried out lawfully except with the approval of the Standing Review and Advisory

Committee. In effect the regulation of experimentation has been handed over to the SRACI with very little restriction on the exercise of its powers. Although the SRACI has to take into account the broad principles in s. 29(7), including s. 29(7)(b) which states that it 'shall ensure that the highest regard is given to the principle that human life shall be preserved and protected at all times', the said human life is not defined. All these omissions in the Act have the effect of leaving these questions to the SRACI.

The functions of the SRACI reflect the role that is envisaged for it. Should this approach be commended as against, for instance, using just legislation? Although some would regard the history, so far, of the SRACI as rather chequered, some of the problems attributed to it spring from the controversy surrounding the nature of its work. Some of its critics confuse the morality of the law and the law itself that the SRACI administers. Considering the area of embryo experimentation in its scientific and ethical perspective, the committee approach is far better than just legislation—if the premise is that legislation is inevitable. The decisions that they have made and recommendations for amendments that incorporate new scientific knowledge are some of the benefits that are offered by such an approach.

Conclusions

In this chapter we have traced the path that has been travelled in Victoria to regulate embryo experimentation by legislation. The committee stage—to make a preliminary inquiry into various implications of the problem—seems to be a near-universal model. Many Australian states and the Commonwealth have appointed committees, as have many other countries.[13] The legislative model, however, is not one that has yet been replicated in many other places.[14]

In the context of the legislation in Victoria, it may well be that further amendments will be required again in future to clarify the application of the Act. In so far as law-making occurs in a context of imperfect knowledge, and is carried on by a varied body of community representatives, it is rare for any piece of legislation to be beyond criticism in all its details. The process of implementing any regulatory scheme through law is dynamic, necessarily so because the values and knowledge present in the community are dynamic. Politicians, lawyers and the community will learn more of the science they are dealing with, and scientists will learn more about the way law and politicians approach regulation and ethical issues are debated and clarified. The Act can be adjusted to suit changing conditions and a more liberal approach may be justified when community fears for the future are allayed. Alternatively, developments in the IVF program may produce calls for stricter legislative control. The Waller Committee and the Victorian Government concentrated on the major issues of principle, with the result that many points of scientific detail were not examined. However, and this is an important point, by 'testing the water' first, the Victorian process has made it easier for those diving in later to avoid some of the snags which have so far been identified in legislating for reproductive technology.

Notes

1 News release issued by the Attorney-General, Victoria, 11 March 1982.

2 These committees and their reports are: Senate Select Committee on the Human Embryo Experimentation Bill 1985, *Human Embryo Experimentation in Australia* (Senator Michael Tate, chairman), (AGPS, Canberra, 1986); Family Law Council: *Creating Children: A Uniform Approach to the Law and Practice of Reproductive Technology in Australia* (AGPS, Canberra, 1985); New South Wales Law Reform Commission: *Report on Human Artificial Insemination* (Sydney, 1986), *Report on In Vitro Fertilization* (Sydney, 1988) and *Report on Surrogate Motherhood* (Sydney, 1988); Queensland: *Report of the Special Committee Appointed by the Queensland Government to Enquire into the Laws Relating to Artificial Insemination, In Vitro Fertilization and Other Related Matters* (Brisbane, 1984); South Australia: *Report of the Select Committee of the Legislative Council on Artificial Insemination by Donor, In Vitro Fertilization and Embryo Transfer Procedures and Related Matters in South Australia* (Parliament House, Adelaide, 1987); Tasmania: Committee to Investigate Artificial Conception and Related Matters, *Final Report* (Hobart, 1985); and Western Australia: *Report of the Committee Appointed by the Western Australian Government to Enquire into the Social, Legal and Ethical Issues Relating to In Vitro Fertilization and its Supervision* (Perth, 1986).

3 For a more general description of the reports, see L. Skene, 'Moral and Legal Issues in the New Biotechnology', *Australian Law Journal* 59 (1985), 379, 385–9.

4 Committee to Consider the Social, Ethical and Legal Issues Arising from In Vitro Fertilization, *Interim Report* (Prof. Louis Waller, chairman), (Victorian Government Printer, 1982), p. 8.

5 See second reading debate on the Bill, Vic. Parl. Debates, 21 March 1984, p. 1935 and 18 April 1984, p. 2317.

6 See ss 9–12 of the Bill introduced in March 1984, ss 10–13 of the Act.

7 Committee to Consider the Social, Ethical and Legal Issues Arising from In Vitro Fertilization, *Report on the Disposition of Embryos Produced by In Vitro Fertilization* (Prof. Louis Waller, chairman), (Victorian Government Printer, 1984), p. 16.

8 This case was resolved in 1987 by a directive from the Minister of Health that the embryos be removed from storage and, if viable, donated anonymously for use in an IVF program: J. Schauble, 'Rios embryos decision welcomed', the *Age* (Melbourne), 4 December 1987.

9 See M. Brumby and P. Kasimba, 'When is cloning lawful?', *Journal of In Vitro Fertilization and Embryo Transfer*, 4 (No. 4, 1987), 198.

10 This was added to the Act by the *Infertility (Medical Procedures) Amendment Act 1987.*

11 Waller, op. cit. (VGPO, August 1984), para. 3.29, p. 47.

12 See S. Buckle, K. Dawson and P. Singer, 'The syngamy debate: When does an embryo begin?', Chapter 19 of this book.

13 See L. Walters, 'Ethics and new reproductive technologies: an international review of committee statements', *Hastings Center Report* Special Supp. 3, 17 (No. 3), (1987), p. 3.

14 In Australia only South Australia has so far followed the Victorian example: see Appendix 1 of this book.

19 | The syngamy debate: When precisely *does a human life* begin?

STEPHEN BUCKLE, KAREN DAWSON
AND PETER SINGER

The *Infertility (Medical Procedures) Act 1984* (Vic.) did not specify precisely when an embryo comes into existence. This may seem a point of merely academic interest; but within two years of the passage of the legislation, it gave rise to a heated controversy with real practical consequences for researchers and infertile couples. The issue was only resolved by the Victorian parliament passing an amendment which in effect specified, more precisely than any legislature had done before, the moment at which egg and sperm become an embryo. This enactment could be seen as a legal declaration of when a life begins—although what legal or ethical significance such a starting point possesses remains an open question.

The significance of the debate reaches beyond the Victorian experience, because it can now be foreseen that it will occur whenever a jurisdiction seeks to prohibit, by statute or other form of regulation, experimentation on human embryos. In this chapter we shall set out the background to the controversy, examine the relevant arguments and describe the way in which the issue was resolved.

The background

The *Infertility (Medical Procedures) Act 1984* (Vic.) made it illegal to fertilize eggs removed from the body of a woman for purposes other than implantation of the resultant embryo in a woman's uterus. In other words, it prohibited the creation of embryos specifically for research purposes.[1] The Act also set up a Standing Review and Advisory Committee on Infertility (henceforth 'the SRACI'), with authority to approve experiments on surplus or 'spare' embryos. Such embryos are originally fertilized with the intention of implanting the resulting embryo in a woman's uterus, but for one reason or another are no longer wanted for that purpose.

In 1986 the SRACI received a proposal to carry out research into the microinjection of a single sperm into the egg, a procedure which would allow subfertile males to become (biological) fathers. The usual requirement for testing an innovation of this kind would be to examine the embryos so produced for genetic and chromosomal abnormalities. This examination requires destroying the embryos concerned. Since the researchers did not consider it ethically defensible to implant the resulting embryos in a woman's uterus until a number of them (40 was the figure suggested in the research proposal) had been examined in this way, the eggs initially removed in order to be fertilized by microinjection would have been removed from the donor's body 'for purposes other than implantation of the resultant embryo'. The SRACI, therefore, could not approve the research, regardless of its merits, since it clearly would contravene the Act.

Since destructive examination of the embryos would be unacceptable, the researchers proposed to carry out a more limited research plan. Monitoring the feasibility of sperm microinjection does not necessarily require that the fertilized egg be allowed to develop to an advanced stage. Whether or not the procedure is reliable can be assessed by the success of the sperm in entering the egg, before the genetic material of the egg and sperm has combined—the stage known as syngamy. Male and female chromosomes can be observed in the pronuclear stage prior to syngamy, and these can also be an indication of normality. Another parameter that can be assessed prior to syngamy is the fertilizability of the egg. This provides a check on whether injecting the sperm under the zona pellucida has damaged the egg. Fertilizability would, in these cases, be determined by establishing whether or not the egg is able to complete its maturation and participate in fertilization.

At this point problems of definition arose. Under the Act, is an egg fertilized as soon as the sperm has entered it? If it is, then testing the egg at this stage is contravening the Act, because it amounts to fertilizing eggs for purposes other than implantation. If, on the other hand, the envelopment of the sperm by the egg is not sufficient for fertilization (or, is not *yet* fertilization), then testing at this stage is within the legal boundaries of the Act. The central problem however, is that the Act fails to specify just what counts as fertilization, and so is unable to settle the question of the legality of the research procedures in question.

In order properly to understand the nature of the issue here, we need to understand the scientist's use of the term 'fertilization', and the reasons for this usage. Briefly, the matter can be put in the following way. As knowledge of the earliest stages of human life has spread, it has become more common to view human fertilization as a complex process, lasting for about 24 hours.[2] The reason for this is that the union of the sperm and egg (the ordinary meaning of 'fertilization' in sexual reproduction) seems to be best identified not with the bare fact of the sperm being incorporated into the egg but with the processes, initiated by the incorporation of the sperm, which culminate in syngamy. In outline, these processes are as follows: the sperm becomes incorporated in the egg, the egg completes maturation, the genetic material of each condenses into chromosomes, and finally the male and female contributions come together to form the new genotype. This first formation of the new genotype is syngamy; and, because the union of the two gametes does not seem to be complete before syngamy has occurred, the proper scientific use of the term 'fertilization' includes the entire process which begins with the sperm passing through the zona pellucida and comes to completion at syngamy.

Given this scientific meaning of 'fertilization', the proposed research did not contravene the Act. But whether the term was properly understood in this way when used in the Act was a problem for the SRACI to resolve. Either it had to settle on a definition of fertilization, or another way of clarifying the issue had to be found. There were no precedents. In other debates on when a human life begins—for instance, in the context of abortion—the fact that the sperm has passed into the egg is not known until long after syngamy has taken place. One might say that the legislators ought to have foreseen the problem; but this is unfair. For all of human history prior to the development of IVF, it had made sense to regard fertilization as an event, rather than a process, because no one had been able to observe it, or to interfere with it between the time the sperm entered the egg and the time when syngamy was complete.

In March 1987 the SRACI discussed whether the research proposal came under the Act and was, therefore, prohibited by it. Opinion was sharply divided; and when a vote was taken, there were four members in favour and four against. The Act did not give the chairman a casting vote. To resolve the dilemma, the SRACI proposed that the Act be amended to allow it to approve research proposals involving the initiation but not the completion of the process of fertilization outside the body of a woman. Although the SRACI was evenly divided on the question of whether the Act as it stood already allowed such research to be approved, a majority of 7–1 supported the idea of amending the Act so as to make it clear that fertilization is a process which is completed at syngamy, and that an embryo begins to exist only when fertilization is complete. This view was the heart of the controversy which the proposal created. Does syngamy really mark the beginning of a new human life?

Against syngamy

As a marker for the beginning of human life, syngamy has its critics. We shall now set out an argument for the view that a human life begins when the sperm first passes into the egg, and not at syngamy. Before directly considering this argument, however, it is important to get one thing clear. The view that a new human life begins at syngamy is not an alternative to the commonly accepted belief that a new human life begins at fertilization. It is, rather, an attempt to explain just how that traditional belief should be understood. Of course, the traditional belief may also be explained in some other way: but it should be stressed that no new standard is being proposed. Our knowledge now shows the traditional belief to be imprecise, so the traditional belief needs to be refined, or interpreted. Syngamy is one such interpretation; and one way of focusing the debate is to ask what other interpretations of the traditional belief are possible.

The view to be considered holds that a new human life has begun once fertilization has begun. In other words, on this view, the new human life exists *before* the completion of the process of fertilization. This can be seen as an attempt to capture the spirit of the traditional belief because it implies an interpretation of why the traditional belief thought fertilization to be important. Since there has been very little serious thought addressed to this issue, we shall take as an expression of this view a submission which was put to the SRACI by the St Vincent's Bioethics Centre.[3] The submission puts its view in this way:

> The [St Vincent's Bioethics] Centre has in the past defended the philosophical opinion that a new human individual comes into existence when the process of the fusion of sperm and ovum results in a single unified entity which is so organized as to have the capacity to develop as the kind of being which would normally have that collection of attributes which we describe as human—especially the ability to doubt, reason, enquire, affirm, understand, love, etc.

The point at issue is when the new human individual begins. The submission takes the view that a new human individual has come into existence as soon as the sperm has entered the egg. The reason given is that:

> when the two membranes [i.e., of sperm and ovum] open to one another and the contents of the sperm are released into the ovum, the sperm loses its separate identity and the ovum gains a capacity it did not have while simply an ovum, that of developing as a human individual . . . The two cells (sperm and ovum) have become a single cell containing many interacting components which by their interaction have the capacity for organizing all the subsequent stages of human development.

There are two grounds offered here for the view that the moment the sperm enters the egg marks the beginning of a new human individual. The first is that this is the moment when the sperm 'loses its separate identity' and (another way of putting the same point) that the two cells have then become 'a single cell', or in the phrase quoted previously, 'a single unified entity'. The second ground is that this is the moment when the ovum, or egg, 'gains

a capacity it did not have . . . that of developing as a human individual
. . .'. We shall refer to the first ground as 'the unification argument' and to
the second as 'the argument from capacity'.

The St Vincent's submission also contained a negative argument,
against attributing any particular importance (biological or moral) to syngamy. The negative argument is, in essence, as follows:

> *Syngamy involves no new chemical process nor a shuffling of genetic
> material, it is little more than the juxta-positioning of the chromosomal
> pairs which already existed and had their places pre-determined before the
> juxta-positioning takes place.*

We shall refer to this simply as 'the argument against syngamy'.

Another argument for the importance of the moment the sperm is
enveloped by the egg should also be mentioned, although since it is discussed
in Chapter 5, it need not be set out in detail. This is the argument from
genetic uniqueness. It can be argued that the significance of the moment
when the sperm passes into the egg is that this is when the unique genetic
identity of the child-to-be is determined. Until that moment, any one of
thousands of sperm could have fertilized the egg, and the resulting child
could have had as many different genetic identities. Once one sperm has
entered (or so the argument runs) the egg 'locks up' so as to prevent other
sperm from entering. Now the die is cast: male or female, fair or dark, the
genetic blueprint can no longer be altered. Syngamy does nothing more
than begin the process of putting into effect what has already been decided
when one particular sperm, rather than the thousands of others, was the
first to pass through the zona pellucida.

The arguments evaluated

The argument against syngamy

We begin with the negative argument, to the conclusion that syngamy
cannot bear the weight attributed to it. At least in the version we have
quoted, this argument fails.

The argument begins with factual claims about what syngamy does *not*
involve: 'Syngamy involves no new chemical process nor a shuffling of
genetic material.' These claims are, at best, imprecise. What is a 'new
chemical process'? On the one hand, it might be said that syngamy is a
chemical process, and it is new in the sense of not having occurred previously
in that particular egg. On the other hand, it could be said that in all of
evolutionary history, there has been no new chemical process since the
emergence of DNA. No doubt the authors have some sense of the term in
mind, but it is not at all clear in what sense the enveloping of the sperm

by the egg involves a 'chemical process' that is 'new', while syngamy does not. Similarly, what is the 'shuffling of genetic material'? Syngamy involves the completion of a process which could be described as the 'shuffling of the genetic material'. Is the completion of a shuffle still part of the shuffle?

Perhaps the factual claims made in the submission could be made more precise and defended. Even if they could, however, the fundamental question is not whether syngamy does or does not involve new chemical processes or the shuffling of genetic material. The real question is when a new individual begins: so, unless there are good reasons for identifying the origin of an individual with the occurrence of a new chemical process, or with a shuffling of genetic material, or with the conjunction of these and other processes, these claims fail to address the point at issue. It is not clear what reasons could be offered. So there is no good reason for believing that the facts offered are relevant. Of course, this does not show that an individual *does* begin at syngamy, only that the facts adduced in the submission might be accurate and yet it remain an open question whether the new individual begins at syngamy.

The unification argument

The unification argument has two sides to it: from the moment that the sperm begins to enter the egg, it is claimed, the sperm loses its identity, and the sperm and egg become a single, unified cell. We shall discuss first the claim that the sperm loses its identity.

It is tempting to think of the egg as if it consisted of a tiny sphere filled with a liquid. On this model, once the sperm has passed through the outer layer of the sphere, it simply dissolves into the liquid inside, losing its identity in the process, as the salt one adds to a bowl of soup loses its identity as it dissolves into the soup. The authors of the submission appear to have this model in mind when they write: '. . . the contents of the head of the sperm are released into the egg cell. The sperm cell ceases to be identifiable at this point . . . its contents mixing with the contents of the ovum.'[4]

The language employed here—of contents being *released*, of *mixing*, of an object *ceasing to be identifiable*—paints a picture of previously discrete objects either dissolving or becoming hopelessly dispersed throughout a fluid. This is most misleading: although the sperm's tail and membrane are absorbed, the genetic material in the head of the sperm remains intact within the egg, and subsequently forms the male pronucleus. Some 20 hours after the sperm has been incorporated into the egg, this pronucleus is a discernible, i.e. identifiable, entity within the cell. It is the merging of this pronucleus with the female pronucleus, beginning approximately 22 hours after the sperm has been incorporated into the egg and taking about two hours, that constitutes syngamy.

Hence it is at least misleading, if not downright false, to say that when the membranes of egg and sperm open to each other, the sperm ceases to be identifiable. Those who do take the loss of identity of the sperm as an important criterion of when a new life begins would be better advised to look towards syngamy, rather than the moment the sperm is enveloped by the egg, as marking this moment.

Once this point is grasped, it is easy to cast doubt on the other side of the unification argument, the claim that once the sperm has been enveloped by the egg, there is a single unified entity. The fact that after sperm envelopment there is only one cell is not a sufficient reason for thinking that there can no longer be two distinct entities. Siamese twins with separate brains can be thought of as two distinct entities contained in a single body: they are physically one single thing, but they are two distinct individuals nonetheless. It is not clear why we should not think of the 'male' pronucleus within the pre-syngamy cell in this way. Why should we not regard the male pronucleus as a distinct entity, albeit wholly contained within another cell?[5]

What we seek here is the first formation of a single *unified* entity. The fact that there is at this stage a single cell does not indicate that the cell has the necessary unity to be regarded as a new individual. Is the pre-syngamy cell unified? This is a difficult question to answer, but not because we do not know enough. The problem is, rather, that it is difficult to determine just what is at stake, because it is difficult to determine just what counts as unification. Nevertheless, there is one sense (at least) in which the pre-syngamy cell is not unified: its genetic constituents are not unified. Genetically speaking, then, it is still in the process of *becoming* unified (and so cannot yet function *as* an individual). This remains true no matter how inevitable the processes might be which will, in time, bring it to be a genetic individual.

The argument from capacity

The problem with the argument from capacity is that it seeks to establish when a new human individual comes to exist, but the argument itself relies on the existence of the *capacity* to develop human characteristics. The problem here is that not only has it not been shown that only distinct individuals possess these capacities but also it seems rather easy to show that the necessary capacities exist before there is a single individual. A motile sperm and an egg, considered as a pair, comprise a system which possesses these capacities *prior to* the sperm's entering the egg. If the system thus identified did not possess at least some of these capacities, fertilization itself could not occur. (If it is objected that we don't know whether or not the sperm will succeed in penetrating the egg—which is, in the normal case, quite true—it can be replied that neither do we know that a sperm-penetrated egg will successfully achieve syngamy, or cell-division, or implantation, etc.

Early life is a hazardous and unpredictable affair—only 30–40% of fertilized eggs survive to birth. The possession of the relevant capacities is no guarantee of successful development, principally because the capacity to become a being of a certain kind is, in the usual case, only a necessary condition for becoming such a being.)

The basic error seems to be the familiar fallacy of affirming the consequent: because a new human individual has the capacities necessary to develop into a fully developed, conscious human being and because a pre-syngamy cell likewise has these capacities, it is erroneously concluded that a pre-syngamy cell must be a new human being. The fallacy can be illustrated as follows: the fact that two entities share a common feature does not show that they are the same—or even that they are similar in any other respect. Cats and dogs are both four-legged; but we cannot infer from this that cats are dogs, nor even that cats are like dogs (even though, in some respects, the latter is true). The fact that pre-syngamy cells have the capacities necessary to develop into conscious human beings, therefore, does not support the conclusion that such cells are already new human beings.

So the argument from capacity is certainly not compelling. We would, however, go further: it is not even plausible. This is because it depends, at crucial points, on misleading or inadequate formulations. This is particularly true of the use made of its central concept: capacity.

It has already been suggested that the argument's concern with capacities involves a shift from its stated concern with determining when a new human life begins, a shift made all the more significant by the likelihood that the relevant capacities exist before the new individual does. If this problem is to be overcome, two strategies might appear to be available. On the one hand, it might be argued that the capacity to develop into a being possessed of the distinctive human attributes (such as the abilities 'to doubt, reason, . . . understand, love, etc.') is itself a human capacity in just the way that these attributes are human capacities. Thus possession of these capacities qualifies a being as a person, because, like all persons, it possesses the capacities necessary to reason, understand, etc. On the other hand, it might be held that, although these capacities are not to be understood simply as personal attributes, nevertheless we should treat the bearers of such capacities in the same way as we treat persons. The first of these strategies attempts to say that the being with the capacities to become a person *is already* a person; while the second is a form of the argument from potential.

Each of these strategies has its problems. The first strategy is based on an equivocation. Its argument attempts to establish that we should treat a new, biologically human, individual in the same way as we treat all human beings—most importantly, we should not interfere with its natural development except to protect its interests; and we should treat them in this way because they have the very same capacities as a human being—the capacities to doubt, reason, enquire, affirm, understand, love, etc. The problem is, however, that when we describe a person as a being with the capacity to understand, or to love, we mean that that individual *can* understand or love.

In the ordinary course of the individual's behaviour we can observe such capacities being *exercised*. With the fertilized egg (whether pre- or post-syngamy), the ascription of such distinctively human capacities is not the ascription of activities in which that individual can now engage; the capacities in this case are not capacities that can be exercised. There are thus two senses in which the term 'capacity' is being used: in the former sense it means what some being can now do; in the latter it means what some being can, in the normal course of events or in appropriately favourable circumstances, come to be able to do. The former kind we can call actual abilities; the latter, potential abilities. The important question to be asked at this point is, thus, *not* 'Does a particular being possess human capacities?' Rather, the important consideration is this: we all recognize the moral importance of actual abilities, as our ordinary moral principles show; but how should we regard merely potential abilities? Can potential abilities justifiably be regarded as of equal importance as actual abilities?

Putting the matter in this way helps to show how exposing the equivocation in the meaning of 'capacity' leads to the argument becoming a form of the argument from potential. The problem here, however, is that there is considerable doubt about whether the argument from potential can succeed. There is no clear sense in which the fertilized egg possesses a potential which is not also possessed by the egg and sperm, when still separate but considered collectively. In the laboratory situation it is not possible to appeal to what the fertilized egg will become if allowed to develop 'naturally', since without human intervention the fertilized egg is bound to die. Nor does it help the argument from potential to derive potentiality from the probability of the fertilized egg becoming a child for, in the laboratory situation, the probability of a newly fertilized egg becoming a child is still very low—even the most experienced teams using *in vitro* fertilization to treat infertility have a success rate below 10% per *embryo* transferred (higher figures quoted refer to the success rate per *transfer*, but each transfer may involve three embryos, and the transfer is considered a success if one child results). Thus the probability of a fertilized egg becoming a child is not significantly greater than the probability of a child resulting from a given egg and some sperm, prior to fertilization. But if potential is not to be understood in terms of a natural course of development, nor in terms of probability, there may be no way in which it can be understood that distinguishes the potential of the fertilized egg from the potential of the egg and sperm, when still separate.[6] So at the very least, much more needs to be done before the authors' employment of considerations of the pre-syngamy cell's capacities can be persuasive.

The argument from genetic identity

The argument from genetic identity, taken as a general argument for fertilization as the moment at which a new life begins, has been discussed in detail in Chapter 5. Here we shall limit ourselves to problems related specifically to the use of the argument to pinpoint the moment the sperm enters the egg as the beginning of the new human life.

The first problem is, again, that the facts are not quite so simple. Recent research has shown that a number of human eggs fertilized *in vitro* have been fertilized by more than one sperm. These eggs have three clusters of chromosomes, or pronuclei, and are therefore referred to as 'tripronuclear'. Estimates of the number of eggs which are tripronuclear vary from 1% to 4%.[7] In these eggs, it is clear that the penetration of the first sperm does *not* determine the genetic identity of any embryo that results. A second sperm may succeed in penetrating the egg as well. This means that the fertilized egg now includes not the usual 46 chromosomes characteristic of *Homo sapiens*, but 69 chromosomes—the egg's original 23, plus 23 from each of the two sperm which have been incorporated into the egg. One cannot even say that the genetic identity of the resulting embryo has been determined at this stage, for a study of 29 tripronuclear fertilized eggs showed that by the two- or three-cell stage of development, some had reverted to a normal 46 chromosomes, whereas others remained at 69, some had a number between 46 and 69, and others had a number outside this range— from 34 to 72! In other words, even after the two sperm have penetrated the egg, some genetic material can be rejected. Even though this phenomenon occurs in only a small minority of fertilizations, the fact that it can occur at all is sufficient to refute the claim that the genetic identity of the embryo is determined at the moment the sperm passes into the egg. For at this moment, the entry of a second sperm is still possible; and even after a second sperm has entered the egg, just which chromosomes will become part of the embryo, and which will be discarded, still appears to be open.[8]

Since we are concerned here with the permissibility of experimentation on embryos *in vitro*, it is scarcely relevant whether these abnormalities occur also in natural reproduction; but in fact they do. One per cent of all human conceptions are embryos with 69 chromosomes ('triploid embryos'). Some of these embryos survive to birth, and although the babies born are severely abnormal, a few have survived for several months.[9]

So here is one reason why the argument from genetic uniqueness does not point to the moment the sperm passes into the egg, rather than syngamy, as the start of a new human life. If there are just a few cases in which another sperm will subsequently be incorporated, and genetic identity will only be fixed at a later stage, then it can never be said with certainty that genetic identity is fixed at the moment the first sperm passes through the zona pellucida. At that particular instant, the possibility of a second sperm doing the same cannot be excluded.[10]

Perhaps current research will very soon require a qualification to the last sentence: but it is a qualification which leads directly to the second reason why the argument from genetic uniqueness cannot accomplish the role it is here being used to play. There is one set of circumstances in which penetration by a second sperm can quite definitely be excluded: when only one sperm can come into contact with the egg. This will be the case if the research proposal on fertilization by microinjection of a single sperm—ironically, the proposal which triggered the Victorian debate on syngamy—succeeds. But the success of this method of fertilization would not make it easier to argue that the moment the sperm penetrates the egg is the beginning of a new life because that is the moment at which the genetic identity of the embryo has been determined. If microinjection of a single sperm ever becomes 100% successful, the genetic identity of the embryo will be determined *before* the sperm penetrates the egg: it will be determined once the sperm has been picked up by the laboratory technician, preparatory to the injection. We would then *either* have to take this moment as the beginning of a new human life, *or* conclude that the determination of genetic identity is not, after all, a sufficient ground for holding that new human life has begun.

The advocate of the argument from genetic unity is unlikely to take the first of these routes, for that would imply that a single sperm, once selected by a technician, cannot be destroyed. At a minimum, some additional condition—such as the existence of a unified entity—will need to be satisfied. Once this is conceded, however, there is room for discussion as to what constitutes unification; and here, as we have seen in considering the argument from capacities, there is reason to hold that the penetration of the sperm does not in itself yield a unified entity.

Resolution by legislation

The arguments considered above were put forward as objections to the proposal that experiments on fertilized eggs be allowed to proceed up to the time of syngamy. The parliament of Victoria apparently did not consider the objections sufficiently convincing, for in November 1987 it passed the *Infertility (Medical Procedures) (Amendment) Act 1987*. The amendment specified that, subject to the approval of the Standing Review and Advisory Committee on Infertility, eggs could be fertilized outside the body for experimental procedures 'from the point of sperm penetration prior to but not including the point of syngamy'. Syngamy itself is defined as 'the alignment on the mitotic spindle of the chromosomes derived from the pronuclei' (ss 4(2) and 4(4)(d)).

After the passage of this amendment, the microinjection research proposals were approved.

Is there an argument for syngamy?

We have seen the weaknesses of three major arguments for the moment the sperm penetrates the egg as a 'marker event' for the beginning of a new human life, and of one argument against syngamy as this marker event. These weaknesses do not show, however, that syngamy deserves the importance attached to it by the amended Victorian legislation. It may be that there are other arguments against regarding syngamy as the beginning of a new human life. In addition to arguments which, like those we have examined, seek to show that a new human life begins *before* syngamy, there may be others which seek to show that a new human individual does not begin to exist until *after* syngamy. An example of the latter kind of argument would be an argument to the effect that a new human individual does not come to exist until the formation of that discrete biological entity which itself will develop into the human fetus. This entity is the embryo proper, which starts to form at about the time the new genetic material begins to implant in the uterus.[11] These questions cannot be considered here; it is sufficient to note that, even if it is accepted that syngamy marks the end of the process of fertilization, it does not follow that syngamy is thereby established as the beginning of a new human life. The traditional view, that a new human life begins at fertilization, is not beyond criticism.

A further complication now needs to be added, especially since the point at issue has been misunderstood by a number of contributors to the IVF debate. Even if it is concluded that syngamy (or any other appropriate event) is the beginning of a new human life, the moral significance of syngamy (or other event) is not thereby established. This is because the settling of factual questions about the beginning of a new biological life does not thereby settle the moral question of how we should treat such biologically defined entities.[12]

Notes

1 For a brief discussion of the basis of this prohibition, see Helga Kuhse and Peter Singer, 'Individuals, humans and persons', Chapter 7 of this book.

2 Harrison, R.G., *Clinical Embryology* (Academic Press, London, 1978), p. 4.

3 The St Vincent's Bioethics Centre is a Victorian centre for research in bioethics which supports Roman Catholic approaches to issues in bioethics. The submission was published anonymously under the title 'Identifying the origin of a human life', in *St Vincent's Bioethics Newsletter* 1987, 5(1), 4–6.

4 Ibid., p. 4.

5 See K. Dawson, 'Segmentation and moral status', Chapter 6 of this book.

6 See P. Singer and K. Dawson, 'IVF technology and the argument from potential' and Stephen Buckle, 'Arguing from potential', chapters 8 and 9 of this book.

7 Mahadevan, M. and Trounson, A., 'The influence of seminal characteristics on the success rate of human *in vitro* fertilization', *Fertility and Sterility*, 42 (1984), 400–5.

8 Kola, I., Trounson, A., Dawson, G. and Rogers, P., 'Tripronuclear human oocytes: altered cleavage patterns and subsequent karyotypic analysis of embryos', *Biology of Reproduction* 37 (1987), 395–401.

9 Jacobs, P. et al., 'The origins of human triploids', *Annals of Human Genetics* 42 (1978), 49–57; Sherard, et al., 'Long survival in a 69XXY triploid male', *American Journal of Medical Genetics* 25 (1986), 307–12. We owe these references to the article by Kola, Trounson, Dawson and Rogers, cited in note 8 above.

10 For a discussion of possible genetic changes after this point, see Karen Dawson, 'Fertilization and moral status', Chapter 5 of this book.

11 See Karen Dawson, 'Fertilization and moral status'; Peter Singer and Karen Dawson, 'IVF technology and the argument from potential'; and Stephen Buckle, 'Arguing from potential', chapters 5, 8 and 9 of this book. See also Norman Ford, *When did I Begin?* (Cambridge University Press, Cambridge, 1988).

12 For one argument that it is not morally important when a new human being begins to exist, see Helga Kuhse and Peter Singer, 'Individuals, humans and persons', Chapter 7 of this book.

APPENDICES

1 | A summary of legislation relating to IVF[1]

The following text relates mainly to Australia, North America and Western Europe. As the situation is changing all the time, the text cannot be accurate to the present time.

Australia

In Australia, legislation to regulate IVF was first passed in Victoria in November 1984. In 1988 and 1991, respectively, South Australia and Western Australia also enacted broad legislation. Although committees of inquiry into IVF and related issues have been held in other states, no attempts have been made to enact legislation solely on IVF. However, all the states and both territories have legislated for the legal status of children conceived by 'artificial' techniques.

New South Wales

The first self-contained Australian statute dealing with the legal status of artificially conceived children was the *Artificial Conception Act 1984* (NSW). It provides that when a husband consents to the use of donor sperm to achieve his wife's pregnancy, he is presumed to have 'caused the pregnancy' and to be the child's father. The sperm donor is presumed *not* to have caused the pregnancy and *not* to be the child's father, whether the recipient woman is married or not. Both these presumptions are irrebuttable. The Act defines 'husband' and 'wife' to include partners of opposite sex living together on a *bona fide* domestic basis.

Victoria

The *Status of Children (Amendment) Act 1984* (Vic.) reaches further than the New South

Wales Act. In addition to creating a presumption of paternity in favour of a consenting husband, the Act covers children born from donated embryos or donated eggs. The Act creates an irrebuttable presumption that the birth mother is the mother of the IVF child and that the ovum donor is irrebuttably *not* the mother.

Victoria was for four years the only state with comprehensive legislation aimed directly at regulating IVF. The *Infertility (Medical Procedures) Act* was enacted in 1984 and amended by the *Infertility (Medical Procedures) (Amendment) Act 1987* (see Appendix 2 for the relevant sections of the Act).

The Act regulates a wide range of IVF activities and includes provisions:
• limiting the practice of IVF to approved hospitals;
• limiting the availability of IVF to married women and laying down other conditions for those who may have access to it;
• providing for record keeping and confidentiality;
• setting up the Standing Review and Advisory Committee to consider requests for approval of embryo research proposals and to advise the minister on infertility; and
• banning commercial surrogacy.

The most controversial of the Act's provisions are those on embryo experimentation.[2] Briefly, the Act:
• permits harmful embryo experiments with the approval of the Standing Review and Advisory Committee;
• bans the fertilization of eggs without the intention to transfer the resultant embryo into a woman;
• permits, with the approval of the committee, research on a fertilizing egg before the stage of syngamy; and
• permits the freezing of ova and embryos in an approved hospital.

South Australia

The *Family Relationships Act 1975* (SA) (as amended in 1984 and 1988) follows the scheme of the Victorian status of children legislation by covering both maternal and paternal presumptions.

In the area of IVF South Australia has, like Victoria, a wide ranging statute. The *Reproductive Technology Act 1988:*[3]
• establishes a Council on Reproductive Technology of 11 members to monitor the practice of IVF and research, and to advise the minister on issues of reproductive technology;
• lays down provisions to guide the council in formulating a code of IVF ethical practice;
• provides for licensing of IVF procedures by the South Australian Health Commission and IVF research by the council;
• prohibits embryo flushing, cryopreservation of embryos beyond 10 years and any research which 'may be detrimental to an embryo'; and
• makes it an offense to divulge confidential information other than as provided in the Act.

Western Australia

The objects of the *Human Reproductive Technology Act 1991* are:

to regulate and guide the use of reproductive technology, by making sure that standards set by the Act are adhered to by licensees; to ensure that artificial fertilization procedures are not used without consideration of the welfare of any child likely to be born as a result, and unless participants have been adequately assessed medically as to the need for these procedures, and are eligible, according to the Act itself; to make sure that general community standards and welfare are considered in any decisions about reproductive technology, and that public debate can be carried out on an informed basis; and to allow beneficial developments in reproductive technology to take place.

The Act establishes the WA Reproductive Technology Council to:

(i) formulate and review a *Code of Practice* to govern the use of artificial fertilization and storage procedures carried on by licensees;

(ii) advise the Commissioner of Health on the suitability of applicants, and compliance of licensees with conditions of their licenses;

(iii) make sure that any research carried out by or on behalf of a licensee, on eggs, sperm or participants, has general or specific approval of Council;

(iv) advise the Minister for Health on matters related to reproductive technology;

(v) encourage and facilitate research into the causes and prevention of all types of human infertility and on the social and public health implications of reproductive technology; and

(vi) promote informed public debate and education on these.

The Bill sets out who must consent to the keeping or use of gametes (eggs and sperm) and embryos, and who has rights of control over them, as follows:

• Gamete providers have rights of control over their own gametes, and must consent to any storage or use of the gametes they have provided until they are used to create an embryo, but they may donate them, in which case these rights are passed on to the recipient.

• Embryos may only be created for implantation into a particular woman. Both members of the couple of whom the woman is a member must consent to any storage or use of an embryo.

  • If the couple disagrees, the rights over the embryo may pass temporarily to the Commissioner of Health until the issue is resolved by Court order or agreement;

  • if one member of the couple dies, rights vest in the survivor;

  • if both die, the Commissioner of Health must authorise that the embryo be allowed to succumb;

  • the embryo may only be donated by the couple to another couple, for the purpose of being implanted into the other woman.

Queensland, Tasmania, the Australian Capital Territory and the Northern Territory

All these jurisdictions have 'status' legislation in the same form as the Victorian one.

These are: *Status of Children Act Amendment Act 1988* (Qld); *Status of Children Amendment Act 1985* (Tas.); *Artificial Conception Act 1985* (WA); Artificial Conception Ordinance 1985 (ACT); and *Status of Children Amendment Act 1985* (NT).

Federal

Status legislation, similar in form to that enacted in Victoria, also exists at the federal level, via the following acts: *Family Act 1975* (Cth) s. 60B, as inserted by the *Family Law Amendment Act 1987* (Cth), and *Marriage Act 1961* (Cth) s. 93, as inserted by the *Marriage Amendment Act 1985* (Cth).

Federal legislative efforts in this area are hampered by jurisdictional problems. Constitutionally, the regulation of medical practice and research lies with the states. The Human Embryo Experimentation Bill 1985, which was introduced in the federal Senate, sought to overcome them and ban destructive non-therapeutic embryo experimentation. However, it did not become law. In 1988 the federal government established the National Bioethics Consultative Committee to advise both the state and federal governments on a wide range of bioethical issues. These go beyond IVF and include genetic engineering and euthanasia.

Austria

"The Draft Federal Law on Procreative Medicine" is on the way to being passed.

Canada

As yet there is no Canadian legislation pertaining to IVF. Only Yukon Territory has a statute, enacted in 1984, on the status of children conceived from artificial insemination. Artificial insemination is defined as including *ex utero* fertilization.[4] Extensive recommendations on IVF have been made by the Ontario Law Reform Commission.[5]

Denmark

In July 1987 the Danish parliament passed legislation establishing a 17-member ethics council for the public health system and biomedical research. The council is charged, among other things, with making recommendations to the minister on draft legislation for 'the protection of fertilized eggs and live pre-embryos and embryos' and to assess the current and future direction of gene therapy and diagnostic techniques on human gametes, pre-embryos and embryos. The Act contains a guiding principle for the council requiring the 'assumption that human life takes its starting point as from the time of conception'.

The Act prohibits cloning, mixing genetically different pre-embryos and embryos and the production of hybrids between human beings and animals. Breach of this provision is punishable with a fine or imprisonment.[6]

France

At present there is no wide-ranging legislation on IVF in France. However, two Decrees passed in 1988 pursuant to Law No. 70-1318 of 1970 (on the reform of the hospital system) relate to the licensing of so-called 'assisted procreation'. Decree No. 88-327 defines the activities involved in assisted procreation as the collection, treatment and transfer of human gametes, and the freezing and transfer of fertilized ova. A license is required to practise these activities; and conditions as to qualifications of staff in participating departments and consultation with the National Commission on Medicine and Reproductive Biology (established by the second Decree—No. 88-328) are laid down for the minister's consideration.[7]

Germany

The law on the Protection of Embryos, drawn up for the federal parliament by the federal Minister of Justice makes it a criminal offence to alter the genetic make-up of human germ cells; to fertilize human ova for research; to do any destructive or damaging embryo research; to engage in sex selection and cloning; and to produce chimeras and hybrids from humans and animals.[8]

Israel

The *Public Health Regulations (In Vitro Fertilization) 1987* were made by the Minister of Health to regulate the practice of IVF. The regulations require that IVF be done in 'recognised department', and lay down conditions for the removal of ova from women, fertilization *in vitro* and freezing of ova and embryos. For example, ova can be removed from a woman only for the purpose of IVF and subsequent transfer into a woman. The woman from whom an ovum is obtained must be undergoing treatment for infertility and her doctor must determine that the removal of her ovum will improve her treatment; and ova and embryos can be frozen for five years with a further five-year extension in specific instances.

Mexico

Regulations were passed in 1986 to implement the health research section of the general law on health. Two provisions may apply to IVF. Section 55 provides that research on embryos (it does not say whether they be IVF or other) will be allowed in accordance with the main law. Section 56 states that 'research on assisted fertilization shall be permissible only if it applied to the resolution of sterility that cannot be resolved in any other manner'.[9]

Netherlands

By Decree of August 1988 (Stb. 379), made under the Law of 1971 on hospital facilities, two items—on laboratories used to produce and store embryos, and on IVF as a treatment for infertility to be licensed by the Minister of Welfare, Public Health and Culture —have been added to existing Decrees. The effect of the amendments is stated to be the establishment of IVF as a special facility and its licensing.[10]

Norway

Law No. 68 of 12 June 1987 on artificial fertilization was passed to regulate the practice of artificial insemination and *in vitro* fertilization. Generally, these procedures can only be carried out in specially approved institutions, and only on married women with the husband's written consent. The legislation permits sperm freezing and storage, and embryo freezing and preservation for not more than 12 months. It prohibits research on fertilized eggs, the donation of gametes, and the freezing of unfertilized eggs. It requires institutions carrying out artificial fertilization to report fully to the Ministry of Social Affairs. It is an offence punishable by a fine or imprisonment to engage in prohibited conduct.[11]

Portugal

Decree-Law No. 319/86 (September 1986) is rather general and provides that 'the collection, manipulation, and conservation of sperm, and any other procedures required by the techniques of artificial human procreation' should be done under the supervision of a physician in authorized public or private facilities.[12]

Spain

Law No. 35/1988 on assisted reproduction procedures is the most detailed law under-taken so far on this subject. It covers artificial insemination, IVF and gamete intra-fallopian tube transfer. It lays down general principles for the application of these procedures that emphasize full disclosure of information, including hazards; informed consent; patient data collection and confidentiality; fertilization of ova for the sole purpose of procreation; and the minimization of spare embryos. It specifies conditions applicable to gamete donors, persons undergoing the procedures, and the status of resultant children. It permits sperm and embryo freezing but prohibits ova freezing until the technique is proven to be safe for thawed ova. Intervention on the *in vitro* and *in utero* embryos is allowed for therapeutic purposes subject to conditions, such as informed consent of the woman involved.

The law also contains several provisions on research and experimentation. Gener-ally, experiments are permitted on gametes provided they are not fertilized for subse-quent implantation. The use of the hamster test to evaluate the fertilization capacity of sperm is allowed up to the two-cell stage of the fertilized egg. Fertilization involving other animal and human gametes is banned unless permission has been obtained from the relevant public authority. *In vitro*, but not *in utero*, embryo experiments, whether for therapeutic or other reasons, are generally permitted subject to conditions laid down in the law. Detailed provisions also spell out several specific situations when *in vitro* embryos can be experimented upon. These range from general research to improve assisted reproduction procedures to cellular ageing and contraception.

Offences are also created by the Act. These include the fertilization of ova other than for procreation, and maintaining embryos beyond 14 days.

Lastly, the law provides for the establishment of the National Commission on Assisted Reproduction, to oversee the provision of the procedures, and, in some cases, to authorize research projects. Membership is to be drawn from the government, profes-sions involved with the procedures, and the public at large. The last group will constitute a council within the commission.[13]

Sweden

Law No. 711 of 14 June 1988 is a short statute on IVF. It provides that a woman's egg can only be fertilized *in vitro* for the purpose of procreation; that before transfer, the woman must be married or in cohabitation; that the partner has consented in writing; that the gametes belong to the couple; and that, unless otherwise authorized, the procedure is done in a general hospital. Breach of the law is punishable by a fine or prison sentence.[14]

Switzerland

The cantons of Basel-Land and Vaud have, respectively, issued a directive and guidelines to regulate IVF. The Basel-Land directive was made pursuant to the 1973 health law, and applies the medico-ethical guidelines of the Swiss Academy of Medical Sciences of 1985 to IVF. It prohibits human genetic manipulation, 'organized' surrogacy and 'trade in and the misuse of embryos for pharmaceutical purposes'.[15]

The Vaud guidelines were issued by the Health Council under a 1985 law on public health. They also apply the guidelines of the Swiss Academy of Medical Sciences, and lay down certain conditions for the practice of IVF. These include: IVF should be used as a last resort by married couples to have children; IVF and embryo transfer must be done only by properly accredited physicians; and these procedures must be practised in a medical establishment equipped with the relevant technical facilities.[16]

United Kingdom

In July 1984 the *Report of the Committee of Inquiry into Human Fertilization and Embryology* (the Warnock Committee) outlined a number of recommendations bearing on IVF. At the time of writing, only one such recommendation has been passed into legislation—the *Surrogacy Arrangements Act 1985*. This Act deals only with surrogacy and bans the use of commercial surrogacy agents. It does not extend to other aspects of IFV.

The Human Fertilization and Embryology Act 1990 brought about the eventual implementation of the Warnock proposals.[17] The Act covers the statutory licensing of clinical IVF, the donation and storage of gametes, and embryo research. While licensed research is permitted, embryos are not allowed to be kept or used after the appearance of the primitive streak (that is not later than 14 days from when eggs and sperm are mixed). The Act also specifically prohibits the placing of an embryo in any animal or the replacing a nucleus of a cell of an embryo with the nucleus from a cell of any other person or embryo or subsequent development of an embryo. The Act does not cover GIFT unless the procedure involves donated gametes. Licensees are bound by the Code of Practice laid down by the HFEA.

The Act specifically amends the Surrogacy Arrangements Act 1985 to make all surrogacy arrangements unenforceable and to extend it to include the use of all new reproductive technologies. The Act also amend the 1967 Abortion Act, reducing the time limit from 28 to 24 weeks of pregnancy; it also brings selective reduction of multiple pregnancies within the Act.

United States of America

Federal legislation governing *in vitro* fertilization has been slow to evolve in the United States. Since the mid 1970s an effective moratorium on the use of federal government funds has been imposed by the Department of Health, Education and Welfare (now

Health and Human Services). In 1979 the department's Ethics Advisory Board examined many of the ethical and legal issues raised by IVF and advised that IVF projects were both acceptable and worthy of federal support.[18] It recommended that a model law be drafted to outline the rights of IVF offspring, parents, physicians and third parties involved in the process. To date the model law is still to be drafted.[19]

At present, there are few state laws that deal expressly with IVF as a means of circumventing infertility. The Illinois *Abortion Act 1977* is the first US statute to directly regulate IVF. It provides that any person who intentionally fertilizes a woman's ovum *ex utero* becomes the custodian of the offspring for the purposes of an 1877 child abuse Act.[20] The legislation gives rise to the possibility that an Illinois physician attempting IVF might have substantial liability. However, a legal challenge to the law prompted an opinion by the state attorney-general that the physician would violate the law if he or she wilfully endangered or injured the conceptus in the pre-implantation stage through abuse or harmful experimentation.[21]

Legislation in Pennsylvania also deals directly with IVF, although it aims at monitoring rather than regulating the procedure. The statute requires the filing of public reports on IVF with the Department of Health—for example, names of all persons conducting or assisting in the fertilization or experimentation process, number of eggs fertilized, number of fertilized eggs destroyed or discarded and number of women implanted with a fertilized egg. Failure to submit the reports incurs a fine.[22]

In Louisiana, an Act passed in July 1986 provides for the definition, capacity, legal status, ownership and the inheritance rights of a human embryo. The Act defines a human embryo as being created from the point of fertilization and prohibits the creation of IVF human embryos solely for research purposes. It vests the IVF embryo with legal personality. Consequently, the embryo has power to sue or be sued in its own right. On the question of the embryo's ownership the Act states: 'An *in vitro* fertilized human ovum is a biological human being which is not the property of the physician...or the donors of the sperm and ovum.' Heavy responsibility is placed on physicians who fertilize ova, to ensure that they keep the embryos safely. The Act also sets out qualifications for places where IVF can take place and who can perform it.[23]

In addition to the laws passed specifically to cover IVF programs, 25 states have enacted laws restricting experimentation on fetuses as a result of the abortion decision of *Roe v. Wade*.[24] Of these states, 15 specifically extend their coverage to research on embryos, although only the New Mexico statute mentions IVF expressly by prohibiting certain kinds of 'clinical research activity' on fetuses. 'Clinical research activity' is to be 'construed liberally to embrace research concerning all physiological processes in man and includes research involving human *in vitro* fertilization'.[25]

Although these fetal research laws were passed before the IVF process was developed (the first IVF clinic at Norfolk, Virginia, did not open until 1979), it has been suggested that the threat of criminal prosecution for research or experimentation on embryos has discouraged physicians from offering IVF services in those states.[26]

Notes

1 Originally based on evidence given by Russell Scott, former Deputy Chairman of the New South Wales Law Reform Commission, to the Senate Select Committee on the

Human Embryo Experimentation Bill 1985. The material has been substantially updated and rewritten by P. Kasimba.

2 These are analysed in more detail in P. Kasimba, 'Experiments on embryos—permissions and prohibitions under the *Infertility (Medical Procedures) Act 1984* (Vic.)', *Australian Law Journal* 60 (1986), 675.

3 See P. Kasimba, 'The South Australian *Reproductive Technology Act 1988*', *Law Institute Journal* 62 (1988), 728.

4 Section 14(1) *Children's Act 1984* (Yukon).

5 Ontario Law Reform Commission, *Human Artificial Reproduction and Related Matters*, Vols I & II (1985).

6 Act No. 353 July 1987. This section has relied on an unofficial translation of the Act by the Danish Center of Human Rights.

7 See *International Digest of Health Legislation* 39(3) (1988) (cited as *Int Dig Hlth Leg*), 645–7.

8 This information was kindly supplied by Mr Ellenberger of the Ministry of Justice, FRG.

9 See *Int Dig Hlth Leg* 38(4) (1987), 791, 796.

10 See *Int Dig Hlth Leg* 40(2) (1989), 400.

11 See *Int Dig Hlth Leg* 38(4) (1987), 788–4.

12 See *Int Dig Hlth Leg* 38(4) (1987), 788.

13 See *Int Dig Hlth Leg* 40(1) (1989), 82–9.

14 See *Int Dig Hlth Leg* 40(1) (1989), 93.

15 See *Int Dig Hlth Leg* 39(1) (1988), 83.

16 See *Int Dig Hlth Leg* 38(1) (1987), 76–7.

17 *Human Fertilization and Embryology: A Framework for Legislation*, Cm 259.

18 Ethics Advisory Board, *HEW Support of Research Involving Human In Vitro Fertilization and Embryo Transfer*, 4 May 1979, pp. 108–12.

19 Andrews, L., *New Conceptions* (St Martin's Press, New York, 1984), pp. 147–8.

20 Ill. Ann. Stat. Ch. 38, Section 81–26 (7) (Smith-Hurd).

21 *Smith v. Hartigan* 556 F. Supp. 157 (N.D.Ill., 1983).

22 Pa. Stat. Ann., Tit. 18, Section 3213(e) (Supp. 1983–1984), cited in Ontario Law Reform Commission, *Human Artificial Reproduction and Related Matters*, Vol II, 1985, pp. 382–3.

23 Act No. 964, 1986 La. Sess. Law Serv. 346 (West).

24 410 US 113 (1973).

25 N.M. Stat. Ann. Section 24-9A-1 (D), cited in Ontario Law Reform Commission report, note 22 above, p. 381.

26 Committee on Science and Technology (US House of Representatives), *Human Embryo Transfer*, 8, 9 August 1984, p. 170.

2 | *Extracts from the* Infertility (Medical Procedures) Act 1984 *(Victoria)*

(as amended, 1987)

Interpretation.

3. (1) In this Act unless the contrary intention appears—

"Approved experimental procedure" means an experimental procedure directly related to the alleviation of infertility and approved by the Standing Review and Advisory Committee in accordance with sections 6(3) and 29(6)(b) and (ba).

"Fertilization procedure" means—

(a) a procedure to which section 10, 11, 12, 13, or 13A applies; or

(b) any other procedure (other than the procedure of artificial insemination) for implanting in the body of a woman—

(i) an ovum produced by that woman or by another woman, whether or not it is fertilized outside the body of the first-mentioned woman; or

(ii) an embryo derived from an ovum produced by that woman or by another woman whether or not it is fertilized outside the body of the first-mentioned woman.

"Syngamy" means the alignment on the mitotic spindle of the chromosomes derived from the pronuclei.

. . .

Procedure not be carried out except in accordance with this Act.

5.   (1) Subject to sub-section (2), a person shall not carry out a fertilization procedure.

Penalty: 100 penalty units or imprisonment for four years.

(2) Sub-section (1) does not apply to a person who carries out a relevant procedure in accordance with this Act.

Prohibition of certain procedures.

6.   (1) A person shall not carry out a prohibited procedure.

Penalty: 100 penalty units or imprisonment for four years.

(2) In sub-section (1), **"prohibited procedure"** means—

> (*a*) cloning; or
>
> (*b*) a procedure under which the gametes of a man or a woman are fertilized by the gametes of an animal.

(3) A person shall not carry out an experimental procedure other than an experimental procedure approved by the Standing Review and Advisory Committee.

Penalty: 100 penalty units or imprisonment for four years.

(4) In sub-section (3), **"experimental procedure"** means a procedure that involves carrying out research on an embryo of a kind that would cause damage to the embryo, would make the embryo unfit for implantation or would reduce the prospects of a pregnancy resulting from the implantation of the embryo.

(5) Where ova are removed from the body of a woman, a person shall not cause or permit fertilisation of any of those ova to commence outside the body of the woman except—

> (*a*) for the purposes of the implantation of embryos derived from those ova in the womb of that woman or another woman in a relevant procedure in accordance with this Act; or
>
> (*b*) for the purposes of a procedure to which section 9A applies that is approved and carried out in accordance with that section.

Penalty: 100 penalty units or imprisonment for four years.

(6) A person shall not carry out a procedure that involves freezing an embryo.

Penalty: 100 penalty units or imprisonment for four years.

(7) Sub-section (6) does not apply to a procedure carried out in an approved hospital that involves freezing an embryo if that procedure is carried

out for the purposes of enabling the embryo to be implanted in the womb of a woman at a later date.

(8) Nothing in this Act prevents or inhibits the carrying out in an approved hospital of research on, and the development of techniques for, freezing or otherwise storing ova removed from the body of a woman . . .

Research on process of fertilisation before syngamy.

9A. (1) A procedure to which this section applies is an experimental procedure involving the fertilisation of a human ovum from the point of sperm penetration prior to but not including the point of syngamy.

(2) A procedure to which this section applies—

(a) must be approved by the Standing Review and Advisory Committee before it is commenced; and

(b) must not be carried out unless—

(i) the ova used in the procedure are the ova of a married woman; and

(ii) the woman and her husband are undergoing, in relation to the carrying out of a fertilisation procedure, examination or treatment of a kind referred to in section 10, 11, 12, 13 or 13A; and

(iii) the woman and her husband have each consented in writing to the use of the woman's ova in a specific approved experimental procedure; and

(iv) a medical practitioner by whom or on whose behalf the procedure is to be carried out is satisfied that the woman and her husband have received counselling in relation to the procedure, including counselling in relation to prescribed matters, from an approved counsellor; and

(v) a medical practitioner by whom or on whose behalf the procedure is to be carried out is satisfied that the carrying out of the procedure is reasonably likely to produce information or establish knowledge indicating procedures (including fertilisation procedures) that might be carried out for the purpose of enabling a woman who has undergone examination or treatment of a kind referred to in section 10, 11, 12, 13 or 13A to become pregnant.

(3) A person must not use semen produced by a man (in this subsection called "the donor") for the purposes of a procedure to which this section applies unless—

(a) the donor and his spouse are undergoing, in relation to the carrying out of a fertilisation procedure, examination or treatment of a kind referred to in section 10, 11, 12, 13 or 13A; and

(b) the donor and (unless he no longer has a spouse) his spouse have each consented in writing to the use of the semen in a specific approved experimental procedure; and

(c) a medical practitioner by whom or on whose behalf the procedure is to be carried out is satisfied that the donor and the spouse (if any) have received counselling in relation to the procedure including counselling in relation to prescribed matters from an approved counsellor; and

(d) a medical practitioner by whom or on whose behalf the procedure is to be carried out is satisfied that the carrying out of the procedure is reasonably likely to produce information or establish knowledge indicating procedures (including fertilisation procedures) that might be carried out for the purpose of enabling a woman who has undergone examination or treatment of a kind referred to in section 10, 11, 12, 13, or 13A to become pregnant.

Penalty: 25 penalty units or imprisonment for one year.

. . .

Standing Review and Advisory Committee.

29. (1) There shall be a Standing Review and Advisory Committee consisting of—

(a) a person holding a qualification in the study of philosophy;

(b) two medical practitioners;

(c) two persons representing religious bodies;

(d) a person qualified in social work;

(e) a legal practitioner; and

(f) a person qualified as a teacher with an interest in community affairs—

appointed by the Minister, one of whom shall be appointed as chairman.

(2) A member of the Committee shall hold office for such period as is specified in the instrument of appointment and shall be eligible for re-appointment.

(3) A member of the Committee may be removed from office at any time by the Minister.

(4) A member of the Committee is not, by reason only of being a member, subject to the *Public Service Act 1974.*

(5) Subject to this section, the Committee may regulate its proceedings in such manner as it thinks fit.

(6) The functions of the Committee are—

    (*a*) to advise the Minister in relation to infertility and procedures for alleviating infertility;

    (*b*) to consider requests for approval of and, if it sees fit, to approve, experimental procedures for the purposes of section 6 (3);

    (*ba*) to consider requests for approval of and, if it sees fit, to approve a procedure to which section 9A applies; and

    (*c*) to advise and report to the Minister on any matters relating to infertility and procedures for alleviating infertility and any other associated matters referred to it by the Minister.

(7) In the exercise of its functions, the Committee—

    (*a*) shall have regard to the principle that childless couples should be assisted in fulfilling their desire to have children;

    (*b*) shall ensure that the highest regard is given to the principle that human life shall be preserved and protected at all times; and

    (*c*) shall have regard to the spirit and intent of the several provisions of this Act.

(8) Where the Committee approves an experimental procedure for the purposes of section 6 (3), or a procedure to which section 9A applies the Committee shall forthwith report the approval to the Minister.

(9) The Committee shall make an annual report to the Minister on—

    (*a*) programmes in Victoria under which relevant procedures were carried out in approved hospitals during the year to which the report relates; and

    (*b*) particulars of each programme carried out in each approved hospital in that year including the number of relevant procedures carried out and the number of participants in each programme.

(10) The Committee may from time to time make such recommendations to the Minister on its activities and on its own operation and composition as it sees fit.

(11) The Committee may collate such information relating to and keep such records of, programmes and procedures to which this Act relates as it sees fit and may collate information relating to, and keep records of, similar programmes and procedures carried out in another State or in a Territory.

(12) The Minister shall cause—

    (a) each report made by the Committee under sub-section (8); and

    (b) each annual report made by the Committee under sub-section (9)—

to be laid before each House of Parliament within 14 sitting days after the Minister receives the report or, if a House of Parliament is not then sitting, within 14 days after the next meeting of that House.

3 | *National Health and Medical Research Council Guidelines*

(Supplementary Note 4 to NH & MRC Statement on Human Experimentation 1983)

IN VITRO FERTILIZATION AND EMBRYO TRANSFER

In vitro fertilisation (IVF) of human ova with human sperm and transfer of the early embryo to the human uterus (embryo transfer, ET) can be a justifiable means of treating infertility. While IVF and ET is an established procedure, much research remains to be done and the NH & MRC Statement on Human Experimentation should continue to apply to all work in this field.

Particular matters that need to be taken into account when ethical aspects are being considered follow.

(1) Every centre or institution offering an IVF and ET program should have all aspects of the program approved by an institutional ethics committee. The institutional ethics committee should ensure that a register is kept of all attempts made at securing pregnancies by these techniques. The register should include details of parentage, the medical aspects of treatment cycles, and a record of success or failure with:

   (i) ovum recovery;
   (ii) fertilization;
   (iii) cleavage;
   (iv) embryo transfer; and
   (v) pregnancy outcome.

These institutional registers, as medical records, should be confidential. Summaries for statistical purposes, including details of any congenital abnormalities among offspring, should be available for collation by a national body.

(2) Although IVF and ET as techniques have an experimental component, the clinical indications for their use, treatment of infertility within an accepted family relationship, are well established. IVF and ET will normally involve the ova and sperm of the partners.

(3) An ovum from a female partner may either be unavailable or unsuitable (e.g. severe genetic disease) for fertilization. In such a situation the following restrictions should apply to ovum donation for embryo transfer to that woman.

   (a) the transfer should be part of treatment within an accepted family relationship;

   (b) the recipient couple should intend to accept the duties and obligations of parenthood;

   (c) consent should be obtained from the donor and the recipient couple;

   (d) there should be no element of commerce between the donor and recipient couple.

(4) A woman could produce a child for an infertile couple from ova and sperm derived from that couple. Because of current inability to determine or define motherhood in this context, this situation is not yet capable of ethical resolution.

(5) Research with sperm, ova or fertilized ova has been and remains inseparable from the development of safe and effective IVF and ET; as part of this research other important scientific information concerning human reproductive biology may emerge. However continuation of embryonic development in vitro beyond the stage at which implantation would normally occur is not acceptable.

(6) Sperm and ova produced for IVF should be considered to belong to the respective donors. The wishes of the donors regarding the use, storage and ultimate disposal of the sperm, ova and resultant embryos should be ascertained and as far as is possible respected by the institution. In the case of the embryos, the donors' joint directions (or the directions of a single surviving donor) should be observed; in the event of disagreement between the donors the institution should be in a position to make decisions.

(7) Storage of human embryos may carry biological and social risks. Storage for transfer should be restricted to early, undifferentiated embryos. Although it may be possible technically to store such embryos indefinitely, time limits for storage should be set in every

case. In defining these time limits account should be taken both of the wishes of the donors and of a set upper limit, which would be of the order of ten years, but which should not be beyond the time of conventional reproductive need or competence of the female donor.

(8) Cloning experiments designed to produce from human tissues viable or potentially viable offspring that are multiple and genetically identical are ethically unacceptable.

(9) In this as in other experimental fields those who conscientiously object to research projects or therapeutic programs conducted by institutions that employ them should not be obliged to participate in those projects or programs to which they object, nor should they be put at a disadvantage because of their objection.

Glossary

COMPILED BY KAREN DAWSON

abortion spontaneous or induced termination of pregnancy.

AI artificial insemination (q.v.).

allele an alternative form of a gene, usually distinguishable by its effect on characteristic, e.g. blue eyes, brown eyes.

amniocentesis a procedure used for prenatal diagnosis at about 14–16 weeks of pregnancy. Fluid from the amniotic sac surrounding the fetus, which also contains fetal cells which have been shed during development, is sampled by a syringe inserted through the uterus. The fluid is used for chromosomal and biochemical testing.

amino acid the building-block of protein.

amnion the membrane that contains the embryo or fetus and amniotic fluid. See also *chorion*.

anaemia condition resulting from a reduced number of red blood cells in the blood.

androgenome a fertilized egg containing two sets of male chromosomes and no female chromosomes; usually arises from dispermic fertilization or fertilization by a diploid sperm and the subsequent exclusion of the female genetic contribution.

antibody a specific protein produced in response to an antigen.

antigen any substance which, when in contact with appropriate tissues in the body, elicits an immune response (i.e. formation of an antibody).

artificial insemination the bringing together of egg and sperm under artificial conditions, usually by introducing sperm into the reproductive tract of the female to allow fertilization to occur normally.

asexual having no sex or without sex.

autonomy literally, 'self-rule', in philosophy 'autonomy' refers to the ability to be self-directing, to govern one's own life.

axon the long extension of a neurone through which nerve impulses travel from one nerve cell to the next.

blastocoele the fluid-filled space of the blastocyst which begins to form about four days after fertilization.

blastocyst a stage in early human development that follows from the formation of the morula. The blastocyst forms about four days after fertilization and is a sphere of cells containing a fluid-filled cavity.

blastomere one of the cells which makes up the blastocyst, produced by cleavage division of the fertilized ovum.

cadaver a dead human body.

cancer an uncontrolled cellular growth which if left unchecked might be fatal.

capacitation the process by which a sperm becomes capable of fertilizing an ovum. *In vivo* this process occurs in the uterus or uterine tubes and is induced by secretions in the female genital tract; *in vitro* it is artificially induced before sperm is added to the fertilization medium.

central nervous system (CNS) that part of the nervous system consisting of the brain and spinal cord.

cervix the narrow lower part of the uterus.

chimera an individual composed of cells derived from at least two zygotes.

chorion the membrane surrounding the amnion, amniotic fluid and the embryo/fetus.

chorionic villus sampling (CVS) a procedure for assessing the chromosomal complement of the embryo/fetus, used between 6–10 weeks after fertilization. The tissue sampled is from tufts of tissue, known as villi, which cover the chorion of the conceptus at this stage of development.

chromosome a thread-like structure in the nucleus of a cell composed chiefly of DNA which makes up the genes. A normal human body cell contains 46 chromosomes, and a human gamete 23 chromosomes.

cleavage division the division of the fertilized egg up until the time of blastocyst formation. The size of the pre-embryo remains unchanged throughout cleavage and the cells produced at each cycle of division become progressively smaller.

cleavage arrest the failure of a pre-embryo to continue through to the next stage of development during cleavage.

cloning asexually deriving separate cells or organisms from a single cell or organism; clones are genetically identical to the cell or organism from which they were derived.

compaction the stage in early development where the blastomeres develop close contact and adhere to each other to form a solid sphere of cells; in humans this usually occurs at about the eight-cell stage.

conceptus the product of conception, including the extra-embryonic membranes.

congenital existing at or before birth.

conjoined twins (also commonly known as Siamese twins) monozygous (identical) twins that have not completely separated during their formation from a single zygote. The attachment may range from superficial, involving only skin, to major, involving sharing of organs.

consequentialism the ethical theory which holds that whether an action is right or wrong is entirely determined by whether it has good or bad consequences.

contraceptive a compound or structure that reduces the likelihood of or prevents conception.

cornea the transparent membrane covering the front of the eye.

cryopreservation storage by freezing. It is used to store IVF pre-embryos and similar methods for ova are being developed.

CVS   chorionic villus sampling (q.v.).

cytogenetics   the study of the chromosomes.

cytoplasm   the contents of a cell excluding the nucleus. It consists of an aqueous solution which contains small, living, membrane-bound structures important in the functioning of the cell known as organelles.

deontology   the area of ethics concerned with obligation and duty, rather than with good and bad; often used to refer to those theories which judge whether an action is right or wrong by whether it conforms to a rule or fulfils a duty. In this sense, a deontological theory is opposed to a consequentialist theory.

differentiation   the process of acquiring individual characteristics, as occurs in progressive diversification of cells and tissues of the developing embryo.

diploid   a cell or tissue having two chromosome sets, as opposed to the haploid situation of gametes which have only one chromosome set.

dispermy   the fertilization of an egg by two sperm instead of one as normally occurs.

DNA (deoxyribonucleic acid)   the chemical that makes up the genetic information in most living species.

donor insemination (DI)   insemination of a woman using sperm obtained from a male donor.

Down's syndrome (also commonly known as 'Mongolism')   a syndrome in humans which is characterized by physiological, behavioural and mental defects; caused by the presence of an extra chromosome 21.

ectogenesis   growth of the human embryo and fetus outside the human body until complete development.

ectopic pregnancy   an extrauterine pregnancy; implantation of the pre-embryo has occurred at some place other than the uterus, such as in the Fallopian tube.

embryo   (1) in popular usage, the fertilized egg and subsequent stages of development; (2) in certain scientific circles, the developing human from about two weeks after fertilization until the end of the eighth week when all major structures are represented. See also *pre-embryo*.

embryo biopsy   the removal of one or two cells from the developing 4–8-cell stage pre-embryo for prenatal diagnosis.

embryogenesis   the process of development of the characteristic form and organs of the body; in humans it occurs from the end of the second week after fertilization until the end of the eighth week.

embryology   study of development of an organism from the time of fertilization of the egg until birth.

embryonic disc   the group of cells from which the embryo will develop, usually visible at the end of the second week of development after fertilization in humans.

embryo transfer (ET)   the introduction of the fertilized egg (pre-embryo) into the uterus.

endometrium   the mucous membrane forming the lining of the uterus. The membrane in which the early embryo implants during pregnancy.

epididymis   a cord-like structure in the testes in which the sperm are stored.

erythrocyte   a red blood cell.

ET   embryo transfer (q.v.).

ethics   (1) loosely used, this term is synonymous with 'morality', and means a set of rules or principles regarding the right way to act; (2) more narrowly used to refer to 'moral philosophy', the area of philosophy concerned with understanding the nature of morality and of right conduct.

extra-embryonic   not part of the embryo proper; usually used in reference to the chorion and amnion.

Fallopian tube   the tube through which the ovum passes from the ovary to the uterus. *In vivo* fertilization occurs in the Fallopian tube.

feminism   a political, social, cultural and ideological movement concerned with overcoming male dominance.

fertilization   the process which renders gametes capable of further development; it begins with contact of the male and female gamete and ends with the formation of the zygote.

fetus   the unborn developing offspring in the post-embryonic period when all major structures have begun to develop; in the human it is the stage from the end of the eighth week of development after fertilization, until birth.

gamete   a reproductive cell, sperm or egg which fuses to form the zygote at fertilization in sexual reproduction.

gamete intra-fallopian transfer   a technique which involves the introduction of sperm and eggs into the Fallopian tubes for fertilization to occur.

gametogenesis   process of formation of the gametes.

gel   a sieve-like chemical matrix used to separate proteins or fragments of DNA on the basis of electrical charge and molecular size.

gene   the basic functional unit of heredity.

gene pool   all the genes in a breeding population.

genetic engineering   the procedures of inserting DNA from animal, plant or bacterial cells into other cells; the DNA transferred is replicated and manifested in the appearance of the cell to which it is transferred. See *germ-line genetic engineering* and *somatic cell genetic engineering* for an explanation of the different kinds of genetic engineering.

gene therapy   replacement of a defective gene in a cell by a normally functional gene. See also *genetic engineering*.

genetics   the study of resemblance between and variation among individuals.

genome   the complete genetic make-up of a gamete or cell.

genotype   the entire genetic constitution of an organism.

germ-line genetic engineering   the introduction of genetic material into sperm, eggs or fertilized eggs. The changes made will be inherited by offspring.

GIFT   gamete intra-fallopian transfer (q.v.).

graft versus host disease (GVHD)   an immune reaction of a transplanted tissue or organ against the host because of immunogenic differences, which may lead to rejection of the tissue or organ.

haemopoiesis   the formation and development of blood cells.

haemorrhage   bleeding, loss of blood.

hormone   a chemical substance produced in the body, by an organ or cells, which has a specific regulatory effect on the activity of a certain organ.

hybrid   the offspring of a cross between two genetically different individuals.

hydatidiform mole   a fleshy mass or tumour in the uterus formed by the degenerative development of a fertilized egg.

immunodeficiency   reduction in immune response, sometimes a complete loss.

immunogenic   producing immunity; evoking an immune response.

implantation   attachment of the blastocyst to the endometrial lining of the uterus and subsequent embedding in the endometrium. Implantation begins at about seven days after fertilization and is complete by about 14 days after fertilization. Also sometimes used to describe transfer of an IVF embryo to a woman's uterus.

infertility   a condition of a couple who are unable to produce children in spite of repeated attempts.

inner cell mass   the centrally located cells in the blastocyst which give rise to the embryo.

insemination   the bringing together of sperm and egg whether naturally or by artificial means.

interspecies   used to describe the offspring resulting from the mating of two different species of organisms, i.e. interspecies hybrids.

in vitro fertilization (lit. 'in glass')   the process of fertilization accomplished outside the body as opposed to the normal situation of *in vivo* fertilization.

in vivo   in the living situation.

IVF   *in vitro* fertilization (q.v.).

karyoplast   the nucleus of a cell, that is a membrane-bound structure which contains chromosomes.

karyotype   the chromosome complement of a cell or organism characterized by the number, size and configuration of the chromosomes.

Klinefelter's syndrome   a human condition due to the presence of at least one extra X-chromosome in the male karyotype, i.e. XXY. The males generally show a decrease in fertility and some feminization.

laparoscopy   a method used for collecting eggs for IVF that involves the insertion of an optical scanner (laparoscope) through a small incision in the abdominal wall; a small tube is also inserted for the removal of the eggs.

Lesch-Nyhan syndrome   a rare human disorder caused by an enzyme deficiency and characterized by physical and mental retardation and self-mutilation.

leukaemia   cancer of the blood.

lymphocyte   a cell capable of producing an antibody in response to a specific antigen and of proliferating to produce a population of such antibody-producing cells.

malignant   cancerous; said of tumours.

meiosis   cell division which occurs during the formation of the gametes, where each daughter cell receives half the number of chromosomes.

menstruation   the cyclic uterine bleeding which occurs at about four-week intervals in the absence of pregnancy during the reproductive life of a female.

messenger RNA   the molecules which transfer genetic information which controls the amino acid sequence of a protein from the DNA to the site of protein synthesis.

metabolism   the sum of all chemical processes in the living organism.

microinjection (of sperm)   see *sperm microinjection*.

mitosis   cell division in the non-reproductive cells of the body in which each daughter cell receives the full chromosome complement.

moral status   in ethics, used normally to refer to the nature of the moral consideration due to a being or thing. For example, to say that a dog has a higher moral status than a rock is to say that there is something about a dog which means that we must give greater consideration to the way we act towards dogs than to the way we act towards rocks; conversely, someone who denied this would be asserting that dogs and rocks have the same moral status.

morula   a solid ball of 12–16 blastomeres that forms at about 3–4 days after fertilization.

mosaic   an individual composed of different groups of cells manifesting a different phenotype.

mutation   a change in the DNA that can affect the genotype and phenotype of an organism, caused by exposure to certain chemicals or occurring spontaneously, i.e. unknown causes.

natural law   an ethical tradition based on the belief that what is right and wrong for all human beings can be discovered by an examination of human nature, powers or capacities.

neurone   a cell of the nervous system that transmits nervous impulses.

NH & MRC   the National Health and Medical Research Council of Australia.

NIH   the American National Institutes of Health.

nucleus   the membrane-bound structure in the cell which contains the chromosomes.

oligospermy   a low number of sperm in the semen.

oncogene   a gene which causes cancer (see also *proto-oncogene*).

oocyte   a female reproductive cell that has not yet completed maturing to form an ovum or egg.

ovum   the female gamete; egg.

parthenogenesis   development of a gamete without fertilization.

PCR   polymerase chain reaction (q.v.).

phenotype   the appearance of a cell or organism which results from an interaction between the genotype and the environment.

placenta   an organ, characteristic of true mammals during pregnancy, which joins the mother and embryo/fetus; it provides nutrition to and removes the waste-products from the embryo/fetus via the mother's blood stream.

pleiotropy   an effect whereby a gene affects two or more apparently unrelated aspects of the phenotype of an organism.

pluralism   in politics, pluralism is the view that society should respect the rights of many different groups with varying religious, ethical and political views, allowing each of them to live as much as possible in accordance with their own beliefs, while not attempting to force these beliefs on other groups with differing views.

pluripotent   able to develop in any one of several ways, such as embryonic cells.

polygenic   used to describe traits which are determined by many genes, each contributing a small effect on the expression of the characteristic.

polymerase chain reaction   a technique used for isolating and repeatedly synthesizing specific sequences of DNA.

polymorphism   the presence of several forms of a gene in a population.

postnatal   occurring after birth.

pre-embryo   alternative name for the 'pre-implantation embryo'; the conceptus from fertilization to the appearance of the primitive streak at about 15 days after fertilization; see also *embryo*.

prenatal   occurring or existing before birth.

prenatal diagnosis   the detection of genetic or developmental abnormalities in the embryo or fetus.

primitive streak   the initial band of cells from which the embryo begins to develop. The primitive streak is present at about 15 days after fertilization.

pronucleus   the egg or sperm nucleus after the incorporation of the sperm into the egg during fertilization.

protein   the compounds which make up many of the structural and functional components of a cell; they are composed basically of amino acids and the sequences are coded for by the DNA of a cell.

proto-oncogene   a gene involved in cell proliferation which may be changed into a cancer-causing gene or oncogene.

radiolabelling   a technique for identifying newly synthesized proteins which involves incubating the DNA or RNA in the presence of amino acids carrying a radioactive component, or label, that will be incorporated in any new protein.

radiosensitive   used to discribe tissues or compounds which respond to or are charged by the presence of radioactivity.

recombinant DNA   DNA that has been artificially introduced into a cell so that it alters the genotype and phenotype of the cell and is replicated along with the native DNA.

reproductive technology   the battery of techniques developed to increase reproductive success: especially IVF and its related procedures.

RNA   (ribonucleic acid)   a molecule transcribed from the DNA of a cell as part of the production of a protein by the cell.

segmentation   division into similar parts, such as in cell division, or the division of the pre-embryo which occurs when identical twins are formed.

semen   the ejaculate of the male which includes the sperm and its nutrient plasma and other secretions.

sentience   strictly, the ability to sense something; but in ethics the term is normally used to refer to the ability to feel (at least) pain.

Siamese twins   see *conjoined twins*.

sociobiology   the attempt to apply our knowledge of biology, and especially of evolutionary theory, to the social behaviour of animals (including humans).

somatic cell   all body cells except for the egg and sperm and the cells from which they develop.

somatic cell gene therapy   the insertion of genetic material into the body cells of an individual. The changes will not be inherited by the offspring.

somatic cell genetic engineering   see *somatic cell gene therapy*.

spare embryo   an embryo produced by IVF that is in excess of the number acceptable for transfer to the woman.

species   group of individuals that interbreed to produce fertile offspring.

sperm   the mature gamete of the male.

sperm microinjection a procedure where a single sperm is injected under the outer covering of the egg for fertilization; may be of value for sub-fertile males whose sperm are weak and unable to penetrate the zona pellucida of the egg.

spindle the structure of fine threads occurring in the cell during cell division to which the chromosomes attach.

SRACI the Standing Review and Advisory Committee on Infertility (in Victoria, Australia).

stem cell a generalized mother cell, the progeny of which may differentiate along several lines and become specialized cells.

sterilization procedure which results in an individual being made incapable of reproduction, e.g. hysterectomy, tubal ligation, vasectomy.

superovulation a treatment resulting in the maturation of many eggs in the one cycle.

surrogacy (or 'surrogate motherhood') term for an arrangement where one woman (the 'surrogate') bears a child for a couple, with the intention that the child is handed over to the couple soon after birth. Usually the surrogate is inseminated with the sperm of the adopting father; in IVF surrogacy gametes from either or both adopting parents may be used to form the pre-embryo before its transfer to the surrogate.

syngamy the final stage in fertilization in which chromosomes from the male and female gametes come together to form the zygote.

thalassaemia a group of inherited disorders affecting the haemoglobin in the red blood cells.

therapeutic used to describe a procedure involved in healing or cure.

tissue culture the growth of living tissue cells in a culture medium conducive to their growth.

tort a civil (i.e. not criminal) wrong in law for which damages can be recovered.

totipotent able to differentiate along any line; for instance, a totipotential cell would be able to give rise to the entire individual with all its cell types.

transcription the transfer of information contained in the sequence of a DNA molecule to a specific RNA molecule which will then be used as a template for the production of a protein.

transgenic used to describe the animal or plant resulting from the introduction of a gene (or genes) from another species which is expressed in the phenotype.

translocation a chromosomal mutation involving a change in the position of a segment of a chromosome.

triple-X female a female containing an additional X-chromosome in the karyotype, i.e. XXX.

triploid a cell, tissue or organism having three chromosome sets (for the human this is 69 chromosomes).

tripronuclear used to describe an egg which contains three pronuclei, usually as the result of fertilization by two sperm.

trophoblast the outer layer of cells surrounding the blastocyst which are involved in implantation by invading the lining of the uterus and later give rise to the extra-embryonic membranes.

trophoblastic disease a condition in pregnancy resulting from over-proliferation of the trophoblast.

tumour a new growth of tissue in which the multiplication of cells is uncontrolled.

Turner's syndrome   a human condition usually associated with the lack of one X-chromosome, i.e. XO. The affected individuals are female, but usually have non-functional or immature sex organs.

ultrasound   the use of high frequency sound waves for visualizing internal structures of the body by reflection of the ultrasound waves.

uterus   the organ of the female in which the pre-embryo becomes embedded and in which the developing embryo and fetus are nourished.

utilitarianism   sometimes used broadly as a synonym for consequentialism (q.v.), classical utilitarianism was the more specific ethical theory that whether an action is right or wrong is entirely determined by whether it produces a greater or lesser nett surplus of pleasure than any alternative action. More recently, other forms of utilitarianism have been proposed, the most notable of which is preference utilitarianism. In this variant, the 'satisfaction of preferences' plays the role filled by 'pleasure' in the classical form of the theory.

vaccine   a suspension injected for the purposes of inducing immunity.

viability   having the capacity to survive independently; with regard to the human fetus viability, in legal terms, is to obtain from 28 weeks of gestation.

womb flushing   the procedure of washing the recently formed embryo from a female's Fallopian tube or uterus before implantation has commenced.

zona pellucida   the outer layer of the egg that persists and surrounds the pre-embryo until about 4–5 days after fertilization.

zygote   the single cell formed by the union of the male and female pronuclei of the gametes after fertilization.

Notes on contributors

Stephen Buckle is Lecturer in Philosophy at the University of Sydney and former Tutor at the Centre for Human Bioethics, Monash University. His book *Natural Law and The Theory of Property: Grotius to Hume* has been published by Oxford University Press.

Max Charlesworth is Emeritus Professor of Philosophy at Deakin University, and a member of the Victorian Government's Standing Review and Advisory Committee on Infertility, the Australian Health Ethics Committee and the Council of the NHMRC. He is also a former member of the Australian National Bioethics Consultative Committee.

Karen Dawson holds a PhD in genetics and has taught in the Department of Genetics at Monash University. She is currently the Information Officer and a Research Fellow at the Centre for Early Human Development and an Honorary Lecturer in the Department of Obstetrics and Gynecology. She was formerly Senior Research Officer at the Centre for Human Bioethics, Monash University.

Beth Gaze studied science and law, and is now a senior lecturer in the Law Faculty at Monash University. Her interests in law include regulation of medical practice and research, and the general interaction between scientific and legal thinking. Among other subjects she has published a book about Civil Liberties, and teaches Law and Discrimination.

Richard Hare, formerly White's Professor of Moral Philosophy at the University of Oxford, is Graduate Research Professor of Philosophy at the University of Florida, Gainesville. His books include *The Language of Morals*, *Freedom and Reason*, *Moral Thinking*, *Plato*, *Essays in Ethical Theory* and *Essays on Political Morality*.

Pascal Kasimba graduated with the degree of Bachelor of Laws from Makerere University, in Uganda, and subsequently obtained a Master of Laws at Monash University. After working as a Research Officer at the Centre for Human Bioethics, Monash University, he is now with the National Companies and Securities Commission.

Helga Kuhse is Director of the Centre for Human Bioethics, Monash University. She is the author of *The Sanctity of Life Doctrine in Medicine: A Critique* and, with Peter Singer, of *Should the Baby Live?* and the co-editor of the journal *Bioethics*.

Peter Singer is Deputy Director of the Centre for Human Bioethics, and was formerly Professor of Philosophy, at Monash University. His books include *Animal Liberation*, Practical Ethics, *The Expanding Circle*, *Marx*, *The Reproduction Revolution* (with Deane Wells) and *Should the Baby Live?* (with Helga Kuhse).

Alan Trounson is Director of the Centre for Early Human Development and Deputy Director of the Institute of Reproduction and Development at Monash University. He headed the scientific side of the Melbourne IVF team which devised the first successful method of embryo freezing, and was responsible for other significant breakthroughs in the development of *in vitro* fertilization.

Mary Anne Warren teaches in the Department of Philosophy at San Francisco State University. She is the author of *Gendercide: The Implications of Sex Selection*.

Index